The Natural Order of Maths

A Manifesto from Basic Addition to the Reality of Negative Volume

Chris Young

The Natural Order of Maths
A Manifesto from Basic Addition to the
Reality of Negative Volume

First published in Australia by Chris Young 2026

A catalogue record for this
book is available from the
National Library of Australia

ISBN: 978-0-646-74458-2 (pbk)

Typesetting and design by Publicious Book Publishing
Published in collaboration with Publicious Book Publishing
www.publicious.com.au

To Tracy, Skye, Willow, and Declan

Thank you for your support, your love,
and for letting me find the way.

This book is dedicated to each of you
with all my love.

Introduction

This journey that you are about to take is not focused on solving problems or offering a typical understanding of how concepts function. It is about revealing the fundamental nature of basic mathematics and the reality of how these concepts truly work. Although many years have been spent researching these terms, the findings and outcomes presented here are the result of deep insight and a commitment to how mathematics was intended to function—not merely how it was invented.

The following information is a demonstration of what is actually occurring within the numbers. I am not concocting a new perspective or forcing a concept to work just to fit a presentation; I am not reaching for answers where they do not exist. Every theory included in this book has been weighed, proven right, and then rigorously tested until it could be proven wrong. If a sub-concept benefited a major idea but contained even a small part that did not work, the entire concept was deemed flawed. This meant discarding the approach completely and starting again from scratch to uncover the correct and intended way the mathematics was supposed to function. What remains in these pages is only that which I have found impossible to prove wrong, no matter how much effort was applied.

A successful foundation in basic mathematics ensures that each chapter leads so naturally to the next that the content becomes predictable before the page is even turned. In this book, no concept is forced to work through clever wording or hidden steps. Mathematics is not a matter of opinion, and it should never be written in a way that tries to convince a reader of something that is not naturally visible. The logic here is plain, direct, and self-evident.

Preface: How to Use the Natural Order

This book is designed to be read in sequence. Because mathematics is a language built on foundations, each chapter provides the 'DNA' for the next.

To the parents sitting at the kitchen table: You do not need a background in advanced mathematics to understand these pages. All you need is a willing mind and, occasionally, a pencil to sketch the physical reality we are discussing.

To the teachers in the classroom: These chapters are provocations. They are designed to spark a "Wait, that actually makes sense" moment in your students. Use the diagram provided for negative integers to illustrate the structural truth of the operation. Use the physical examples of the hole and the shovel. Consider the physical relationship between a hole (the negative) and the shovel (the operation); the ground level represents the zero boundary.

The goal is not the memorisation of rules; the goal is the observation of the universe.

Contents

1

Subtraction, Minus and Negative, What is the Difference?

Mathematics has three concepts that often exhibit conflicting qualities for the new and advanced learners. While these ideas seem distinct at first, they eventually interweave, creating a web of complexity that can feel daunting to navigate. This challenge is magnified by a lack of consensus on how these principles should be taught. When interpretations vary, it creates a ripple effect where a single concept takes on multiple, often obscure meanings, forcing the learner to re-learn the same idea over and over. While the academic debate over these differences has persisted for years, this book aims to silence the noise and reveal the underlying simplicity that connects them all. Adding to the complexity, two of these concepts—subtraction operators and negative signs—are often used interchangeably, when each should have its own distinct position and significance. In order to discern the variances and commonalities between these three concepts, a significant amount of time has been dedicated to analysing them. This involved delving into numerous mathematical textbooks and published papers, exploring online educational platforms, consulting dictionaries, and engaging in discussions within online educational and mathematical forums. What was the outcome for these three seemingly simple concepts? Subtraction itself is fairly straightforward. For the most part, the majority of people concur that this concept has one meaning, though there are still ideas about subtraction that cause other concepts to be misunderstood. Minus and negative, on the other hand, are where the interesting discussions begin. Occasionally, these two ideas may serve the same objective, while in other instances; they complement each other with distinct definitions. The circumstances under which they are applied can shift abruptly as individuals opt to employ either concept in their preferred manner. This variability arises due to the numerous interpretations of mathematics. Although maths is evidently comprehensible, the evolution of different concepts and the introduction of various shortcuts have somewhat diluted the clarity and precision of mathematics when utilised at a fundamental level.

$$5 + 7 + (-6)$$

This expression is merely an illustration used to determine the underlying factors that contribute to misunderstandings. This expression has an addition operator followed by what appears to be a negative sign, $+(-6)$. In reality, the operator and the sign of the number are positioned as addition and subtraction symbols. Some individuals might not identify a problem with this, whereas others could reach a certain level of uncertainty.

Another concept to examine is how subtraction, minus, and negative are interchanged as calculations progress within an algorithm. This situation may arise when the subtraction sign is positioned in an abstract manner. For instance, when employing an algebraic algorithm to determine the value of a variable, the subtraction operator seems to change its meaning during the process of rearranging the equation. It becomes particularly challenging when a learner observes a certain approach while solving a problem, but then encounters difficulties when attempting to apply the same method to a different concept. It is crucial to examine the mechanisms of each idea in order to understand how they work and the absolute need to clarify their intended meaning to help prevent possible misinterpretations.

What is Subtraction?

The concept of subtraction is to take an amount away from an established group, to remove a certain value, or to find the difference between two values—even from zero. This concept is what is known as an operation, which means it exists within an expression or equation to provide information on what is to happen to the following number. The subtraction symbol, which looks like a dash –, tells the reader exactly what action to perform. This operation instructs the reader to subtract the next value, whether that value is negative or positive. Unfortunately, the dash itself has the ability to be misunderstood, misused, or mistaken for either a subtraction operator or a negative sign. However, the concept is incapable of performing two functions simultaneously; it should not be regarded as representing two distinct states of existence. It is solely an act of subtraction. The most important idea of all is that the subtraction symbol is not a negative. For example, in a simple expression displaying how 2 is being subtracted from 6:

$$6 - 2$$

Given that the six in this expression is evidently positive, the subtraction of two can be misleading, as the value of the 2 is sometimes incorrectly seen as negative. It is essential to comprehend that every number within this specific expression is positive.

Basic addition encounters challenges similar to those of subtraction. In the following example:

$$6 + 3$$

The addition symbol does not make the three positive; rather, it is the absence of a negative sign that indicates the three is positive. In a perfectly clear world, it would look like this:

$$^{+}6 + {}^{+}3$$

While this notation is not novel, the fact that it includes the superscript positive sign facilitates enhanced visual clarity, promoting better understanding for those new to the foundational principles of mathematics.

What is Negative?

In order to precisely understand the concept of negative, it is logical to begin by identifying the purpose and function of the term. While negative is often thought of as simply being a negative number, that idea can be obscure; a negative sign may be present without resulting in a negative value. To elaborate on the previous discussion: subtraction is an operator used for working out problems, yet both the subtraction operator and the negative sign are traditionally shown as a dash placed centrally before a number or variable. This consistent placement poses a significant challenge for learners trying to differentiate between the two. Moreover, the absence of clear guidelines regarding the placement of these distinct concepts creates potential for misunderstanding.

In the natural world and the world of mathematics, all objects, values, and quantities are either positive or negative. Entities considered negative are consistently associated with negative signs in mathematical writing to ensure constant awareness of their numerical values. On the other hand, the positive sign is rarely encountered in written form, except when being taught to beginners, at which point it quickly disappears as another mathematical shortcut. Even though positive numbers usually lack a superscript sign, they serve exactly the same purpose as numbers with a negative sign: both describe the state or value of a result. The term final result refers to any number found after calculations have concluded or one derived from a previous calculation (i.e., a positive or negative amount). Furthermore, positive and negative signs are not operators; they inform the reader of the value of a number before, during, and after calculations. These signs work in conjunction with operators to communicate the ongoing process throughout a calculation.

As stated earlier, a negative sign does not always result in a negative total. For example, in calculus, a negative sign is used in exponential decay formulas to describe a quantity decreasing at a constant rate—such as hot water cooling to room temperature. The negative sign is used with the exponent to allow the calculations to show decreasing values, but the result itself may not necessarily be a negative number. In these instances, the term negative is much more accurate than minus, yet many people use minus to describe results and may never use the term negative at all.

The statement that two negatives make a positive is a useful shortcut, but it has documented issues and can, at times, be incorrect. This occurs when a subtraction operator is mislabelled as a negative. At no point does the negative sign of a number mean subtraction; a negative number must be either added or subtracted using an operator and cannot be calculated without one. To show the difference between a negative number and the operation of subtraction, this book will show a number being negative by using what is known as a superscript minus sign:

$$^{-}5 - {}^{-}3$$

A superscript positive sign ($^{+}$) will be inserted before a number for clarity when needed; otherwise, the number will always be assumed to be positive if one is not included.

What is Minus?

The word minus is a term with multiple, often conflicting meanings. It is commonly used to describe the action of subtraction e.g.

$$3 - 2$$

This can be read as "three minus two". It can also be seen as the identity of a negative value e.g.

$$^{-}4 + 1$$

This is read as minus four plus one. Because this language is used interchangeably in everyday speech, it frequently creates an opening for misunderstanding for the learner. To ensure absolute clarity throughout this book, the term minus will be used to signify the action of subtraction only. The term negative will be used exclusively to describe the identity of a value. This distinction is vital because negative represents a single, consistent concept, whereas minus is often stretched across two different contexts, diluting its mathematical precision.

A final observation must be made regarding the physical placement of the minus operator versus the negative identity sign as they appear in traditional notation. While not always made explicit, the use of white space is often used as a critical indicator of intent; a space between the minus sign and the following number can signify the action of subtraction as seen in the following example:

$$5 - 2$$

However, the absence of a space indicates a negative identity e.g. -4. While these common typographical distinctions will be referenced when discussing current techniques, this book will ultimately implement a more precise superscript notation with no white space to ensure the identity of a negative value is never confused with the action of subtraction; the subtraction action itself will always be explicitly separated by white space.

2

The Very Basic Workings of Addition and Subtraction

In an academic setting, students are initially taught addition and subtraction using physical objects to help them understand those concepts. This approach is very effective because the tangible resources used are relevant to real-life situations. As students become more skilled at mental calculations and rely less on physical objects, the importance of applying addition and subtraction to real-world math problems appeared to decrease significantly. However, it is promising to see that educational methods are now including more real-life examples than ever before.

Looking at the next most basic concept after counting is learned, is the act of joining two groups of things together, or removing some things from another group to create a different amount—and yes, also to discover the concept of difference. Addition and subtraction are the first two basic concepts that perform calculations of multiple amounts or numbers. Even though the processes of these two are quite simple, examples of how they really work are going to become a little more complicated. This is due to the fact that a comprehensive understanding of these concepts is crucial for the remaining majority of this book. It is not that most individuals lack comprehension of addition and subtraction, but rather that a significant amount of information is omitted during the instruction of these two concepts.

When the journey of mathematics begins and addition and subtraction are introduced, young learners are given lots of practice questions which provide simple examples. These may include a basic mathematical story, such as: "Ashwin is holding two apples and is given five more apples; how many apples does she have altogether?" An expression for this may look like the following:

$$2 + 5$$

It needs to be remembered that there is a difference between the operators and signed numbers, as this is crucial in comprehending and understanding how more advanced mathematical processes work. While keeping this in mind, it may not be obvious, but there is more to this expression than meets the eye. For instance, where did the apples come from? The person who is holding two apples clearly got them from somewhere because, at some point in time, they were holding nothing to begin with. When performing a mathematical calculation, it is very important to understand that all addition and subtraction calculations start from zero. The reason for this is that objects to be counted do not just magically appear; they are objects in the real world and everything has a beginning and an end. This small but important concept is only for producing greater clarity when looking at all the concepts throughout this book; addition and subtraction can, of course, be done at any time without the need to physically start from zero. The information given here will shape the expression to look like the following example:

$$0 + 2 + 5$$

This concept of starting from zero is important because it allows learners to understand the mathematical story with greater clarity. It also allows for greater comprehension of the fact that advanced mathematical calculations start from zero as well. In fact, there are three concepts preceding the number 2 that were observed in the expression above. Two of them have already been shown: a zero and an addition operator. There is one more that is not shown, and that is the positive sign:

$$0 + {}^{+}2 + {}^{+}5$$

Proficient students naturally understand that the number two is being added as well as being a positive integer. They may also be aware, without consciously considering it, that basic addition and subtraction operations begin from zero. Therefore, it is crucial to introduce novice learners to these three fundamental concepts that are often overlooked during the initial stages of learning.

The following expression is an illustration of a basic subtraction that any novice student could practice. After discovering how operators and signed numbers work together, learners are presented with a new strategy which is called adding a negative:

$$5 - 2$$

The conversion of a subtraction problem into an addition problem was established for several specific reasons. Foremost among these is the unification of mathematical rules; by treating all operations as addition, mathematicians gain the use of communicative and associative properties which subtraction does not naturally possess. Second, this method was designed to handle the conceptual framework of debt, where the act of removing a debt is equated to gaining a positive asset. The following example demonstrates how this is typically presented to a learner:

$$5 + (-2)$$

Once again, the negative two is written as a subtraction of two, but the concept is supposed to be a negative two. Rewriting the expression will give it greater clarity and understanding for the learner:

$$5 \; + \; ({}^{-}2)$$

As discovered in Chapter 1, a negative does not subtract; it is there as a description of the answer to a math problem or information in a mathematical story to be used within a mathematical calculation. Teaching that subtraction can be changed to adding a negative is a concrete step, teaching how subtraction decreases as calculations move left on the number line. To continue using this concept further into later years of mathematics, it starts to reveal a few problems:

1. In the expression:

$$5 - 2$$

The 2 is never a negative number. It is a positive value. Subtraction is already complex, and turning the 2 into a negative number introduces unnecessary confusion. Consider a real-life situation: “five apples minus two apples.” The traditional system often forces a shift to adding a negative instruction:

$$5 + (-2)$$

Read as a real-world story, this now means: "five apples add two negative apples." Straight away, a problem emerges—negative apples do not exist in physical reality. Unless there is extra context defining the apples as a debt or a loss, the story is incorrect. While the system teaches this shift to prepare learners for advanced negative numbers, teaching basic learners to simply take away the apples is all that is required.

2. Another associated issue is when learners start adding numbers together; they are told that addition moves in a positive direction, which is to the right on the number line. When adding a negative on the positive side of the number line, it now adds to the left, which may cause some misconceptions.

3. The more complicated issue is when learners become more advanced; they encounter an addition operator and a negative sign together. Learners are taught to change this into a subtraction problem; this, of course, contradicts what they had previously learnt, which in turn may complicate subtraction even further.

When learners are taught about adding a negative by using the standard definition:

$$a - b = a + (-b)$$

It becomes clear that using larger numbers will once again pose a challenge.

This can be seen in action if some numbers are substituted:

$$473 - 259 = 473 + (-259)$$

To work this out mathematically, the addition column algorithm would be called upon, which would then have to be set up as below:

$$\begin{array}{r} 473 \\ + -259 \end{array}$$

Learners do not know how to use the addition algorithm in this manner; as a matter of fact, no one does. Subtraction is not converted and it is always done as the following:

$$\begin{array}{r} 473 \\ -\ 259 \end{array}$$

If learners were to encounter an addition operator and a negative sign, they can change this to being a subtraction expression. If they encounter a subtraction symbol, then they still know what to do. While the standard definition of adding a negative may appear structurally sound, the simplicity found in low-value calculations often obscures underlying problems that cannot be ignored.

The complexity of this concept becomes particularly evident when applied to algebra, where the adding a negative concept results in an unnecessarily complicated notation:

$$4x + (-5x) + 4y + (-2x) + (-2y)$$

This is far less complicated when seen as it originally was:

$$4x - 5x + 4y - 2x - 2y$$

The concept of adding a negative aligns with the traditional definition of the additive inverse, where an inverse value is added to the original number, results in zero:

$$6 + (-6) = 0$$

While the actual definition simply states 'a number added', this phrasing contributes to the obscurity between the act of addition and the state of the negative value. The added number is referring to the negative six being added to the positive six, which calculates it back to zero. This concept is useful for more advanced calculations such as algebra, trigonometry, and more. The additive inverse will be looked at in greater depth later in Chapter 9.

Inaccurate negative or subtraction explanations in algebra can lead learners to develop a false sense of confidence regarding the interpretation of numerical sentences.

The next example is not from a mathematical story; it is a random calculation created to illustrate the obscurity between subtraction and negative concepts:

$$3 - 2x = 9$$

This could be read as "three minus two *x* equals nine." When solving this algebraic equation for x, the process may start by subtracting three from both sides of the equation, leaving the next lines of the equation to work on:

$$3 - 3 - 2x = 9 - 3$$
$$-2x = 6$$

Reading this as "minus two x equals six" creates potential for error in subsequent steps. The original equation utilised minus to indicate the subtraction of positive $2x$; however, the $2x$ now appears as a negative identity. The critical question remains: is the symbol a subtraction operator or a negative sign? To resolve this, a basic test will be conducted. If the $2x$ is a negative identity, it requires a preceding operator. Converting the number sentence to include an addition operator provides the necessary verification:

$$3 - 2x = 9$$
$$3 + {}^{-}2x = 9$$
$$3 - 3 + {}^{-}2x = 9 - 3$$
$$+ {}^{-}2x = 6$$
$$x = {}^{-}3$$

Although this works very well, it still has the issue of changing what appears to be a positive $2x$ into a negative $2x$, which is not ideal as there is no mathematical instruction to be able to do this. Towards the end of the calculations, it also dismisses the addition operator situated before the negative $2x$, but this may be acceptable because it can be added to zero and not need stating. The second way of testing is to keep the $2x$ as positive because it is being subtracted, while also adding some missing symbols and zeros to see how the equation behaves:

$$0 + {}^{+}3 - {}^{+}2x = {}^{+}9$$
$$0 + {}^{+}3 - {}^{+}3 - {}^{+}2x = {}^{+}9 - {}^{+}3$$
$$0 - {}^{+}2x = {}^{+}6$$

Of course, equations are not dealt with in this way, but this discussion is not about how to solve equations; it is about seeing what the values and their signs actually are. The last line above shows that the calculations evolve to subtracting a positive multiplication; it is unclear how to proceed and appears that this method may not work. Upon delving a little deeper into how this could work naturally, it was discovered that if the last line on the LHS was evaluated, the calculations could proceed as normal:

$$0 - {}^{+}2x = {}^{+}6$$

Literally subtract ${}^{+}2x$ from 0 to become a negative value:

$$^{-}2x = 6$$
$$x = {}^{-}3$$

The addition operator in the first test worked as intended; however, the issues raised did not offer enough evidence to establish its accuracy. The second test shows how the real calculations are performed and why the subtraction operator appears to magically transform into a negative when calculated using basic algebra.

In conclusion, it is clear to see that $2x$ is in fact a positive value, but that term does end up being negative. Although this is not through an out of the blue decision to just change it, the change occurs simply by adhering to a mathematical process. The other concept that has been clearly shown is that the subtraction operator is unequivocally an operator and not a negative at any time. Even towards the end of the calculations, the subtraction operator remains subtracting until the terms are evaluated. The subtraction operator disappears simply because the operation of performing subtraction has taken place, and that operation can only happen once.

One of the least explainable concepts in basic mathematics is subtracting a negative. It appears that uncovering a genuine explanation is rather challenging; however, numerous methods have been developed to demonstrate how to accomplish this concept. Commonly taught methods include walking forward then backward, turning to walk in reverse, or cancelling out chips. None of these provide viable reasoning for the underlying mechanics of what is actually occurring. If an example is shown that includes the use of arrows and pointers, it creates the perception that its accuracy is based on its achievability, so it is not the best way to teach this concept.

Subtracting a negative can be explained very easily even without a number line, but a number line will be used nonetheless to help those who are visual learners. The following example shows the necessary information to highlight exactly what is occurring within the expression:

$$0 + {}^{-}7 - {}^{-}4$$

Having the calculations start from zero shows that seven negative objects or entities have been added to zero. Showing this ensures that negative seven exists as real objects on the number line. Now, if four of those real objects are literally taken away, it will leave three negative objects. This is not a ploy or any form of deception. It is achieved without resorting to misleading or unclear language. The explanation is straightforward as it accurately reflects what is happening.

The hard part of understanding how to subtract a negative is the way that it is normally shown on a number line. If there are negative values less than negative seven on the number line (i.e., negative eight, negative nine and so on), it seems a little difficult to take some of those values away. This is because it is very rarely done in basic maths, but if the same concept is done on the positive side of the number with seven minus four and extra numbers included (i.e., eight, nine and so on), subtracting four is fully understood and acceptable.

Another way to look at it is to consider only the numbers up to and including the large minuend (the larger number) (see fig.1.).

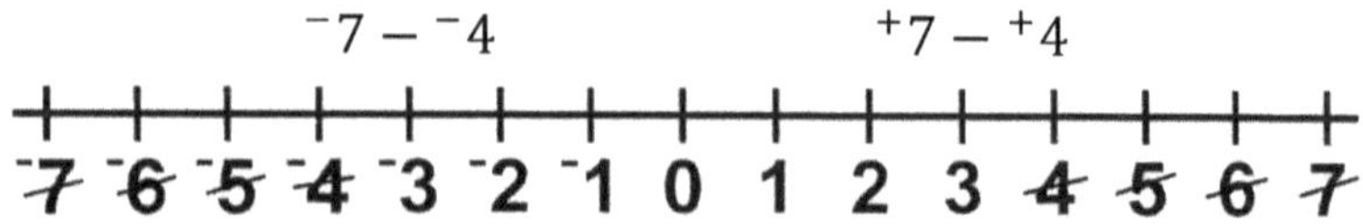

Fig.1. Subtracting a negative

Looking at the image above, it can be seen exactly how $^{-}4$ is subtracted from $^{-}7$, one unit at a time. By restricting the number line to show only the value up to $^{-}7$, the step-by-step reduction becomes clear. This is perfectly logical, as all numbers less than $^{-}7$ on the number line always exist but only in a theoretical state.

3

Commutation and What This Means for Subtraction

The concept of commutation illustrates to students that addition problems can be calculated in different orders yet still produce the same outcomes. This can potentially simplify mathematics for beginners. It is generally considered that subtraction does not follow the commutative property, as it leads to an inaccurate solution when attempted. While these concepts are not ground breaking, a closer examination can reveal underlying patterns and potential challenges. A simple example of addition:

$$5 + 4 = 9$$

With all the extra information added:

$$0 + {}^{+}5 + {}^{+}4 = 9$$

Can easily be commuted to:

$$0 + {}^{+}4 + {}^{+}5 = 9$$

This is straightforward; while it shows how commutation works on the surface, it merely indicates that the positive signs appear to commute, though this is not yet structurally proven. Even if the movement of the numbers alongside their signs seems obvious, what may not be as evident to the reader is that the addition operators themselves remain completely stationary and do not commute. This crucial distinction reveals exactly why commutation with subtraction fails to work the same way as addition:

$$0 + {}^{+}5 - {}^{+}4 = 1$$

If the numbers are commuted, a different answer will result:

$$0 + {}^{+}4 - {}^{+}5 = {}^{-}1$$

It is important to note that the issue does not stem from the numbers, but rather from the operators. Commutation fulfils its intended purpose precisely; however, the primary drawback of learning commutation is a lack of contextual information.

Mathematics is often carried out without specific context, particularly when engaging in repetitive tasks to grasp a fresh concept. This is typically not an issue, as repetition is an effective method of comprehension. However, as an expression evolves during commutation, it is essential to acknowledge that the resulting expression must have come from a completely new mathematical story. When considering apples as an example, obtaining one set of four apples and one set of five apples, combining them can be done in any order. This could have been achieved by adding five apples and then four or four apples and then five. Where commutation can become disorienting is when, for example, making a fruit salad. Adding five pieces of apple and then four pieces of banana in that order cannot have the numbers switched, as this would imply a different starting intent. Commutativity is fine when learning simple addition, but it fails the moment real-world context is introduced. Another issue, albeit a small one, is when there are multiple subtraction operations in a row. Commutation with subtraction cannot be done, yet in situations like the following, it works exactly the same as addition:

$$12 - 4 - 5 = 3$$

$$12 - 5 - 4 = 3$$

This is a simple idea, but there is a bigger issue. When learners become more advanced, they begin subtracting numbers using a column algorithm. Everything works wonderfully until a slight change in value throws everything into chaos. This happens when learners first encounter a larger number being subtracted from a smaller number. Keeping in mind that learners are taught that they *cannot* commute a number that is being subtracted, they are then told that they *can*. For instance, to calculate $241 - 723$, learners are taught to identify the largest number and rearrange the problem placing that larger number on top:

$$\begin{array}{r} 723 \\ -241 \end{array}$$

If this example is placed back into a straight-line expression, it becomes eerily familiar:

$$723 - 241$$

This result is exactly what learners were taught not to do. From start to finish:

$$241 - 723 \text{ turns into } 723 - 241$$

Hence, learners are taught to commute subtraction when it is convenient. Commuting numbers like this technically cannot be done, but it is taught anyway because it seems there is no other easy way. The method of turning the above expression into a column algorithm is trying to accomplish a task that is actually quite simple. In Chapter 6, it will be shown the precise way to subtract a larger number from a smaller number without guesswork or using absolute values.

Rearranging vs. Commutation

The following is an in-depth exploration of a more sophisticated approach: allowing numbers to move around during calculations. This is called rearranging an expression. Starting with a simple addition problem:

$$3 + 4$$

Add the important hidden information and rearrange:

$$0 + {}^{+}3 + {}^{+}\mathbf{4} = {}^{+}7$$

$$0 + {}^{+}\mathbf{4} + {}^{+}3 = {}^{+}7$$

What is happening here is that the addition operator appears to stay stationary, but the positive sign, that is bound to the four at all times, moves with the four; likewise for the three. When a number is placed in an expression, it is always being added or subtracted regardless of its position. The process of subtraction using the rearranging method is straightforward:

$$0 + {}^{+}3 - {}^{+}\mathbf{4} = {}^{-}1$$

$$0 - {}^{+}\mathbf{4} + {}^{-}3 = {}^{-}1$$

This is exactly the same as above, but the subtraction operator has moved with the positive four, resulting in the same answer. The subtraction operator is not a negative sign, but the answer results in a negative value.

Rearranging larger equations is more intuitive when terms are left as their original values—avoiding the clutter of adding a negative. Observe how the logic remains intact during rearrangement compared to the errors often found in standard commutation:

$$0 + 3 - 5 + 7 - 2 + 4$$

Rearranging	Commutation
$0 + 3 - 5 + 7 - 2 + 4$	$0 + 3 - 5 + 7 - 2 + 4$
$= 0 + 3 + 7 + 4 - 5 - 2$	$\neq 0 + 3 - 7 + 4 - 5 + 2$

Grouping like operators makes calculations simple. A mathematical expression can be rearranged in any order, provided the operator moves with its corresponding number—a crucial distinction that standard commutation fails to address.

In algebra, rearranging is already taught using the phrase "gather like terms":

$$8x - 4y + 6 - 3x - 1 + 7y$$

$$8x - 3x + 6 - 1 + 7y - 4y$$

$$5x + 3y + 5$$

No commutative properties are utilised here; the terms are simply being rearranged according to their specific identities. This is also observed in mental mathematics:

$$27 - 33 + 73 + 83$$

$$(27 + 73) + (83 - 33)$$

$$100 + 50$$

Associative Properties

The associative property allows values to be grouped in different combinations without changing their order:

$$2 + 3 + 4$$

$$(2 + 3) + 4 = 2 + (3 + 4)$$

Standard theory claims subtraction is non-associative because the results are inconsistent. The following examples demonstrate this unpredictability:

$$9 - 3 + 5 \rightarrow (9 - 3) + 5 \neq 9 - (3 + 5)$$

$$4 + 2 - 1 \rightarrow (4 + 2) - 1 = 4 + (2 - 1)$$

The first example shows how the regrouping has different results, which of course is expected. However, if the addition and subtraction operators are switched around, the result of the different grouping can work perfectly. This familiar information serves as a basis for analysing why the associative property is not consistently effective for subtraction and for seeking a more efficient or natural method for executing the regrouping process.

Previous statements in this book have indicated that a number that is being subtracted is always being subtracted, no matter what is being done before or after it; this of course is exactly the same for addition. This understanding is significant because it highlights exactly why the standard associative process fails: it tries to move a number while ignoring the fixed instruction to subtract it. Upon further inspection of the aforementioned expression, it is clear that the subtraction operator has been disassociated from the number three in the second grouping:

$$(9-3)+5 \neq 9-(3+5)$$

This changes the expression on the right hand side and shows that the number three is not being subtracted anymore. If the regrouping is done a little differently, which is to allow the three to be subtracted, another error will occur:

$$(9-3)+5 \neq 9\,(-3+5)$$

The issue that has arisen is the absence of an operator between the number nine and the parentheses that would signify multiplication, which is clearly not the intended outcome. However, if the principle is embraced that multiplication is simply an accelerated form of addition, it stands to reason that an addition operator can be implied in that space just as easily as a multiplication symbol. If a multiplication symbol is permitted to exist without being written, then logically, an addition operator can also be present but invisible, surfacing only when needed to maintain the correct relationship between terms. This concept will prompt questioning as to what would happen if a subtraction operator is moved to a different location; could this present an opportunity to fill the empty space with a primary operator, as in the situation above. This theory will be examined to determine its validity across the following calculations:

$$9+3+5=17$$

$$(9+3)+5=12+5=17$$

$$9+(3+5)=9+8=17$$

Concentrating on the three plus five within the parentheses on the third line, it appears to be basic arithmetic with no hidden concepts. From earlier findings, it was found that zero can be inserted at the start of addition and subtraction expressions.

This may give some indication of what is actually occurring while performing the associative properties on the following example:

$$0 + 9 + (0 + 3 + 5) = 0 + 9 + 8$$

The zero and addition operator before the nine can be deleted as it does not serve any further purposes. However, the zero and addition operator before the three is more important and shows what is really happening inside the parentheses. Now that the hidden concept of starting from zero has been included, there is now an extra addition operator which is a perfectly mathematical inevitability:

$$9 + 3 + 5$$

$$9 + (0 + 3 + 5)$$

The problem that has now arisen is that the addition operator from the first line appears to have moved into one of two possible places, as can be seen above on the second line after the 9 and before the 3. An expression containing only addition operators and positive numbers provides a false sense of security; it cannot be used to validate a concept because the results will remain consistent regardless of how the terms are grouped. By changing just one idea in the expression, it can help improve the clarity of the concept being addressed. If the subtraction operator stays with the nine on the outside of the parentheses, it will cause an error as seen in the first trial below:

$$9 - 3 + 5 = 11$$

$$(0 + 9 - 3) + 5 = 11$$

$$9 - (0 + 3 + 5) = 1$$

The second trial will show three still being subtracted over both steps:

$$9 - 3 + 5 = 11$$

$$(0 + 9 - 3) + 5 = 11$$

$$9 + (0 - 3 + 5) = 11$$

The second trial also demonstrates that this logic is correct. After this single change of an addition operator for the three to a subtraction operator, it clearly shows that the addition operator outside of the parentheses does appear there. The subtraction of three must always stay as subtraction, so the subtraction operator must move into the parentheses with the three. This means that the appearing addition operator is always present, but only visible as needed. As previously mentioned, it was anticipated that this outcome would occur and is not too difficult to see how the association property can now include subtraction.

In this laboratory, there are no addition operators to provide a false sense of security. The validity of this concept is further evidenced when examining an expression containing only subtraction operators, such as:

$$9 - 4 - 3$$

$$(0 + 9 - 4) - 3 = 9 + (0 - 4 - 3)$$

This confirms that the appearing addition operator is always present, maintaining the correct relationship even when the primary instruction is exclusively to subtract. Naturally, the same structural reality dictates pure addition. When the associative property is applied to addition alone, an addition symbol operates outside the parentheses in the exact same manner. While standard mathematical shortcuts typically conceal this mechanism, the underlying process is identical.

This information is not presented as a proof; moreover, this material is gained from simple logic of reasoning coupled with explicit knowledge of multiplication being the exact same concept as addition, just a faster or more powerful operator.

$$8 - 3 + 5$$

$$(8 - 3) + 5 = 8 + (-3 + 5)$$

After discovering that an addition operator can be inserted before the parentheses, a search was undertaken to find sources to substantiate this; none could be verified. To test this theory, many calculations were made across various concepts, large, small, positive, and negative values with no errors occurring. This concept is not to be confused with adding a negative, as the numbers are always being subtracted.

4

Fractions vs Division

Numerous studies explore why fractions challenge educational settings. This chapter, however, does not offer pedagogical alternatives; instead, it investigates the fundamental mechanics and global application of these concepts. The primary focus is to uncover the exact disparities between fractions and division to determine which holds greater structural significance.

Fractions originated thousands of years ago, preceding decimals—historically termed decimal inches—by many centuries. While history is not the primary focus that fractions preceded division is essential to understanding their modern relationship. The following comparison traces how each concept functions from foundational levels to advanced applications.

Fractions	Division
Identity: A fraction represents a part of a whole and is expressed as a rational number. It can function as a negative, an exponent, or a collection of similar objects. **Structure:** A numerator is divided by a denominator. It	**Action:** Division is one of the four primary operators (addition, subtraction, multiplication, and division). Each operator functions to create a new value from a mathematical narrative. **Process:** Division partitions a

separates a single whole into a designated quantity of pieces.	number into equal parts containing equal amounts. It can also separate a whole into a set amount of pieces.
Basic Fraction $\frac{3}{4}$  Three parts of the whole.	**Basic Division** $4 \div 2$ or 4/2 or $\frac{4}{2}$ 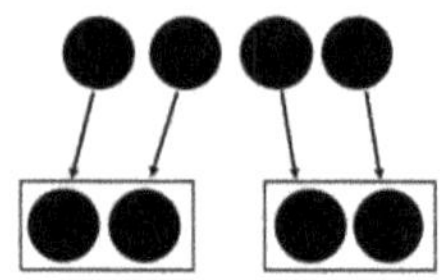
Next Level Fractions  $\frac{3}{7}$ Three sevenths of the circles are black.	**Next Level Division** $12 \div 4$ $12 - 4 - 4 - 4 = 3$ groups.
Words for Fractions $\frac{\text{Numerator}}{\text{Denominator}}$	**Words for Division** $\frac{\text{Dividend}}{\text{Divisor}}$
Vinculum Means the expression may be thought of as grouped together. A unity. Possibly division. Another name, fraction bar.	**Vinculum** Means to divide. Possibly a unity. Other names, fraction bar, division bar.

Types of Fractions Proper Fraction $\frac{3}{4}$ Mixed Fraction $2\frac{3}{4}$ Improper fraction $\frac{4}{3}$	**Types of Division** Quotative division Partitive division Repeated subtraction Multiplication related facts
Harder Fractions $\frac{7}{91}$	**Harder Division** $\frac{7}{91}$
Vinculum, More Advanced Fractions Means to divide	**Vinculum, More Advanced Division** Means to divide
Difficult Fractions Algebra $\frac{2a + 3b\ -\ c}{4a - b + 3d}$	**Difficult Division Algebra** $\frac{2a + 3b\ -\ c}{4a - b + 3d}$
Higher Levels Fractions Calculus $Y + \frac{dy}{dx} = 7$	**Higher Level Division Calculus** $Y + \frac{dy}{dx} = 7$
Fraction Completed Answers $\frac{3}{\sqrt{2}}$ Left as division e.g. trigonometry. 0.75 After dividing a fraction.	**Division Completed Answers** $\frac{3}{\sqrt{2}}$ Left as division e.g. trigonometry. 0.75 After performing division.

The details above showcase the distinct nature of fractions and division in the beginning, but as they evolve into more complex procedures, they are eventually considered to be the same.

How Do Fractions and Division Really Work and Are Their Processes the Same?

When these concepts are taught, learners are usually shown their similarities and differences, but the transition back and forth between them remains extremely complex. To discover how these two ideas truly function, they must be looked at from the moment they are first created and their processes followed until a final result is achieved. The line between two numbers is often called a fraction bar, a division bar, or a vinculum. These terms are used interchangeably, yet none appear to have a universally agreed-upon definition. To avoid confusion throughout this chapter, only the term vinculum will be employed. Because this term already serves as the foundation line in basic arithmetic algorithms, using it here prevents the learner from falsely assuming that every horizontal line must indicate a division. The mechanical function of fractions and division can then be described without distraction from what the vinculum might traditionally imply in any specific instance.

The Numerator and the Dividend

The numerator of a fraction and the dividend of a division problem are frequently treated as the same entity. In many educational settings, these terms are used interchangeably, suggesting they share a similar meaning or function. However, others argue for unique definitions, using each term exclusively within its own domain. These differing interpretations create a landscape of conflicting answers regarding their actual differences, practical uses, and structural significance.

The following two examples are specifically designed to mirror each other, highlighting how easily the similarities between fractions and division can be misinterpreted.

The Fraction Story: "Barry drew a circle and then a line straight down the middle. He then coloured in one side of the circle. What fraction of the circle did he colour in?"

Because Barry coloured one piece out of the two total pieces, the result is expressed as:

$$1/2$$

The Division Story: "Barry drew a circle and then a line straight down the middle. He then coloured in one side of the circle. How much of the circle did he colour in?"

In this instance, Barry began with one circle and then identified one equal part of that circle. The mathematical action is:

$$1 \div 2 = 0.5$$

At first glance, the mathematical processes appear nearly identical. Both stories begin with the same object—a single circle—but this is where the commonality ends. To obtain a Numerator, one must first understand how the Denominator was acquired or created. The numerator cannot exist without the prior partitioning of the whole. In contrast, the Dividend of division requires no such prior creation; it exists as an independent, unaltered whole before any action is taken.

The Denominator and the Divisor

When discussing simple proper fractions, the denominator is always derived from a single one whole and is never expressed as a decimal. In a fraction, the bottom number under the vinculum represents this whole. It is pulled directly from the descriptive information in the mathematical story—such as Barry and his circle. This information is the key to establishing exactly how many parts the one whole has been divided into; that total is then placed under the vinculum to define the denominator. Even if a story involves several objects, such as six pencils, that entire collection is treated as the one whole and becomes the denominator.

A fascinating piece of information often hidden in fractions is what can be called the invisible 1. When a circle is divided into three pieces, the numerical value of that original one whole is not actually stated in the fraction itself. Instead, the fraction only displays the *parts* of that whole. Unless the denominator is 1, the original 1 stays behind the scenes, existing only in the story. In this state, the 1 has been structurally replaced by its partitions. Even when the denominator is 1, it is still strictly a representation of a part of the whole, born from the initial partitioning described in the story.

Finding and understanding the denominator can be complex, but identifying the divisor in a division problem is relatively straightforward. Like the denominator, the divisor is found by searching the mathematical story for specific keywords. Words such as groups, bags of, or each usually signal the divisor. For example, if a story mentions nine bags or sharing with five people, those numbers are destined for the bottom of the division problem.

However, a major disparity exists here: the number stated as the divisor is placed under the vinculum and can be a decimal value—something a fraction denominator can never be.

In division, the 1 on top (the dividend) has not disappeared; it is waiting to be acted upon by the divisor. While the same keywords might be used to describe a denominator, the mechanical rules are different; a divisor is a count of groups, whereas a denominator is a definition of parts that have already replaced the whole.

Fractions and Division: Creation and Operation

The following image illustrates the sequential steps of creating a fraction and a division problem, with a detailed analysis of each step provided below the figure.

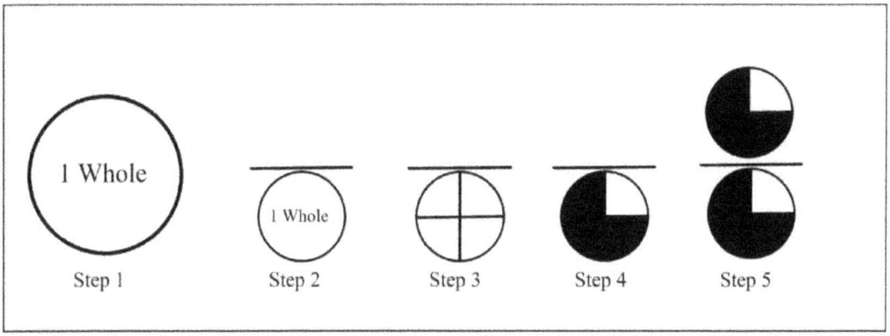

Fig.2. Creating a fraction step by step

Fraction Step 1

This first step in creating a fraction is identifying the one whole, and every numerator is obtained from the one whole. The one whole can and is often more than one object but is still one whole, see Steps for Fraction with multiple objects below. The following fraction steps will use one object as the one whole. When exploring a mathematical story to obtain the one whole and then dividing it into parts, it must be remembered that the one whole is always an integer and never a decimal. When first introduced to fractions, one object is usually represented by a circular, rectangular or square shape, sometimes a pizza or a coloured disc is used, and can be many different objects.

Fraction Step 2

This particular step holds great significance in terms of understanding how fractions evolve. This is because partitioning is used to divide the one whole into a set of equal parts. This idea is far from original, yet the manner in which it is understood and carried out remains hidden. The question then becomes: Why is it that the total number of equal parts—the denominator—is always a whole integer? When creating a fraction, the way that one whole is divided into equal parts is different from the way that division is used to divide an amount of something into a set of groups. For clarity, the term partitioning will be used for the fraction steps and the term division will be used for the division steps, because mathematically, a division instruction does not perform a partitioning act.

The difference between the two concepts, is quite subtle, but the processes used and the final outcomes achieved are significantly contrasting. For a fraction, the one whole is partitioned into a predetermined amount of pieces, which is not a predetermined amount of groups. Furthermore, the one whole is partitioned into equal sizes but this kind of division process allows for those pieces to be slightly different dimensions when needed. This is most evident when partitioning a single object; it demonstrates why decimals or remainders are absent when defining these equal parts. For example, if a mathematical story requires an apple to be partitioned into three pieces, it is done easily because a fraction is not concerned with exactness of size.

Since an apple is not symmetrical, such a partition is acceptable for a fraction, yet it is never performed this way for real division. This principle also applies to a circle; it can be partitioned into an odd number of amounts. Once again, the fraction does not require numerical exactness, provided there is an accurate count of pieces and each piece is relatively close in size. When the one whole is identified, it is placed under the vinculum before any partitioning is done to it, see Fig.2.

Fraction Step 3

The one whole that has been partitioned into the desired amount of pieces which has been taken from the mathematical story can now be placed under the vinculum. This partitioned amount is still one whole and will always remain to be one whole. The obtained amount is now on the bottom of the fraction and is called the denominator. At this point it is not known which of the pieces will be the centre of attention and the denominator will remain as it is and will not change. Taking a look at the image Fig.2, it is clearly visible that there is a circle that has been partitioned into four equal parts, and there are no shaded areas. This idea is consistently represented in numerical form; furthermore, the figure above remains a vital visual aid to help learners understand what is actually happening.

Fraction Step 4

This step is a part of making a fraction that is never seen, and is a concept that does not really need to be seen. What is trying to be understood here is what is actually happening when a fraction is created. So it is essential that this step is seen in comparison of how division works. When finding the actual amount of what the mathematical story is indicating i.e. how many parts of the whole is the main focus, the numerator can be established. After finding out the necessary information, it is indicated by highlighting them in the parts of the one whole that is currently under the vinculum, which of course is the denominator. What is being seen is what would normally be seen as a visual representation of the one whole divided into four pieces, see Fig.2. The difference here of course is where it is positioned under the vinculum. It does look a little strange being here, but this is what is actually happening, and looking as strange as it does is why it is not normally seen at this step.

Fraction Step 5

The second last step in creating a fraction is the most vital of all, simply because this gives the answer and completes the fraction. The filled in parts that are in the denominator are literally copied to the top of the fraction to become the numerator, see Fig.2. It is very obvious but it is still worthwhile mentioning that the filled in parts that can be seen in the top and the bottom of the fraction remain where they are. Furthermore, the filled in parts, top and bottom, are parts of the same thing, they are not independent entities as they all come from the one whole.

Fraction Step 6

The concluding step, after the fraction has been completely created, is attempting to gain a decimal answer, which requires performing real division. This involves taking the three parts of the numerator and dividing it back into the four parts of the denominator. This operation is physically impossible. Since the numerator is a carbon copy of three of the parts in the denominator, division in this context is an attempt to divide a quantity back into itself. The four pieces of the denominator are partitions of a single unit, not independent groups, and the numerator is identical to three of those pieces.

Division Step 1

The first step is to identify the total quantity of the dividend. This value represents the starting whole and can encompass any numerical form, from simple integers to complex irrational numbers. For this demonstration, a single unit is utilised as the one whole to maintain a clear and functional starting point.

Division Step 2

This one whole regardless of its specific value is never divided into smaller parts at this stage. Instead, it is placed above the vinculum as the entire, unaltered dividend. This placement is fundamentally different from that of a fraction, where the one whole is defined below the vinculum. In division, the whole remains intact as the subject of the operation; in a fraction, the whole has already been structurally redefined by the partition.

Division Step 3

The divisor is determined from the mathematical story. This value represents the number of groups the dividend will be distributed into. Because the divisor is an instructional command rather than a structural definition, it can be any value, including decimals or irrational numbers.

Division Step 4

The amount of groups is placed under the vinculum as a numerical value. This is not considered a whole amount; each group is a single group in its own right. These number values can be anything from natural numbers to irrational numbers.

Division Step 5

The dividend is divided into exact, equal parts as determined by the divisor, and these parts are then distributed into their respective groups. When the dividend cannot be distributed into these groups exactly, the remaining amounts are further divided, resulting in either a terminating or a non-terminating decimal. This decimal represents the precise measurement of the parts that were required to complete the division process.

Fraction with Multiple Objects Step 1

The initial stage for fractions with multiple objects is identical to the first fraction step previously established. The distinction lies in the quantity of objects comprising the one whole, which is derived directly from the mathematical story. In an educational context, this fraction is constructed using various discrete objects, such as five apples or seven pencils, to represent the entirety of the system.

Fraction with Multiple Objects Step 2

This stage mirrors the second fraction step, with the exception that multiple distinct pieces constitute the whole. Partitioning is unnecessary; the process requires only a tally of the total pieces present. Whether the objects are different shapes, sizes, or colours, they remain parts of a single whole. Just as partitioned pieces of a single object may be uneven; these multiple objects do not require symmetry or uniformity to be valid components under the vinculum.

Fraction with Multiple Objects Step 3

Since the one whole does not require partitioning, the total number of objects tallied is placed under the vinculum as the denominator. Even though the system consists of multiple whole pieces, they are combined to represent one whole unit. At this stage, no objects are highlighted, as the specific focus of the mathematical story has not yet been determined.

Fraction with Multiple Objects Step 4

This step involves the placement of a value below the vinculum. For instance, if a set contains seven pencils and the mathematical story identifies three as the focus, those three are designated as the active components of the fraction. This process demonstrates the construction of a fraction using multiple independent objects rather than sections of a single item.

Fraction with Multiple Objects Step 5

The quantity of highlighted objects from the denominator is identified and placed above the vinculum as the numerator. This completes the fraction. There is no subsequent step for division, as these discrete objects cannot be divided back into themselves within this specific framework.

Summary

When attempting to convert a fraction into a decimal, several structural problems emerge. First, the numerator—which would theoretically function as a dividend—must consist of an independent whole or multiple whole pieces. In a fraction, this is physically impossible, as every segment of the Numerator is inherently a sub-part of the original whole defined by the denominator.

Second, the denominator would need to function as a divisor (predetermined groups), yet its actual identity is a single whole partitioned into smaller segments. A structural definition of a part cannot be substituted for an active command to divide.

Following the established steps for division, it becomes evident that the Fraction architecture is incompatible with the division process.

Because fractions were developed long before Hindu-Arabic numerals and the subsequent emergence of decimals, they were never intended to produce a whole or decimal result. Historically and structurally, a fraction is not a numerical value; it is a symbolic representation—a static answer to a question of proportion.

In the Real World: The Fraction Mask

To understand mathematics, it is necessary to recognise that two different tools are used that happen to look identical on the page. In the laboratory of reality, a distinction exists between a physical partition and an active instruction.

1. The Physical Fraction (Kitchen Logic)

In everyday life the kitchen, the workshop, or the marketplace a fraction is a state of being. It describes an original whole that has been frozen into equal parts. A measurement of $1/3$ of a cup of flour is a finished, physical object. This is Kitchen Logic, where the goal is to maintain the integrity of the unit. In this realm will always equal exactly 1:

$$1/3 + 1/3 + 1/3$$

Whether the unit is being reassembled, that had been partitioned, or constructing a new unit from measured parts, the logic is unshakeable: the parts define the whole, and the whole is the sum of its identities.

2. The Conflict of Merged Identities (The ballistic formula)

The primary hiccup in mathematical education occurs because these two distinct species are presented as one. By failing to distinguish between them, the system forces students to apply Kitchen Logic to a Ballistic Formula. When a variable or a radical appears in a fraction, the brain searches for a physical part that does not exist. Recognising the symbol as a mask for an active process protects the logic of basic arithmetic while preparing the learner for the precision of higher mathematics.

5

Subtraction of Multiple Numbers

When learners start subtracting more than one number it presents with a somewhat difficult challenge. There are two basic techniques that can be used to work out the subtraction of multiple numbers from a larger number. These techniques are quite easy to understand and simple to utilise. What is required is a method that is a bit simpler and takes less time to calculate. Before delving into a new technique for subtracting multiple numbers, the two existing methods are introduced for the purpose of comparison.

Here is an example that is not too difficult:

$$8324 - 1846 - 932 - 2725$$

1. The first method is to calculate, one part of the problem at a time until each step is completed and then arriving at a solution. This method is done by using subtraction from start to finish:

$$8324 - 1846 - 932 - 2725$$

$$8324 - 1846 = 6478$$
$$6478 - 932 = 5546$$
$$5546 - 2725 = 2821$$

2. The next method is very simple as well but is about addition and subtraction being used together. Before starting, it is necessary to estimate that the minuend is greater than the rest

of the subtrahends combined or it will not work properly. First, identify the subtrahends, which are $1846, 932$ and 2725, then add them together. Second, subtract this result from the first number, 8324 which is called the minuend:

$$8324 - 1846 - 932 - 2725$$

$$1846 + 932 + 2725 = 5503$$
$$8324 - 5503 = 2821$$

This method is more efficient; however, there is a way to obtain the answer through a more direct application of the numerical identities involved in the problem.

After gaining a full understanding of how addition and subtraction function together, another way to solve these calculations was discovered—one that is not completely new, but simply not taught. This method follows a natural process which is very simple to understand and implement. The following is not the best or only way of doing this; it is just a simple solution to a problem that has been around for countless years.

Borrowing powers of ten when subtracting is a mechanical habit that is often used as a default. This means that when needed for subtraction learners can borrow tens, hundreds, thousands and so on. What has been discovered now is being able to borrow more in a single go, such as borrowing twenty or maybe thirty for example when more is needed when subtracting units. The following is like combining the previous two strategies, where adding is still needed and then using subtraction to finish:

$$7945 - 1597 - 284 - 3678$$

$$\begin{array}{r} 7945 \\ 1597 \\ 284 \\ \underline{-\ 3678} \end{array}$$

The process that is about to be shown is a little different from what people doing maths are used to, but seeing and trying it one or two times, makes it relatively easy to do after that. The only concept that is truly different; is borrowing more than ten or one hundred from the number directly to the left. The other concept to keep in mind is that the subtraction operator means to subtract the three lower numbers from the top number. This setup idea is not new; it has always been set up this way to display the subtraction. When starting a problem like this, the equation is taken apart exactly like the previous tasks. Before beginning the steps below to subtract multiple numbers, it is vital to estimate whether the total being subtracted is less than the starting top number. If the estimate is incorrect—meaning the top number is smaller than the sum of the numbers below—the result will be negative. As a negative result would be incorrect for this specific process, this scenario is addressed using a different method in Chapter 6.

$$7945 - 1597 - 284 - 3678$$

$$\begin{array}{r} 7945 \\ 1597 \\ 284 \\ \underline{-\ 3678} \\ 6 \end{array}$$

For the units, add the $7, 4$ and 8

$$7 + 4 + 8 = 19$$

$5 - 19$ will not work.

Borrow 2 from the tens to become greater than 19.

$$20 + 5 = 25$$

$$25 - 19 = 6$$

$$\begin{array}{r} 7925 \\ 1597 \\ 284 \\ -\underline{3678} \\ 86 \end{array}$$

For the tens, add the $9, 8$ and 7 $(90, 80$ and $70)$

$$9 + 8 + 7 = 24\ (240)$$

$2 - 24$ will not work.

Borrow 3 from the hundreds to become greater than 24.

$$30 + 2 = 32\ (300 + 20 = 320)$$
$$32 - 24 = 8\ (320 - 240 = 80)$$

$$\begin{array}{r} 7625 \\ 1597 \\ 284 \\ -\underline{3678} \\ 386 \end{array}$$

For the hundreds, add the $5, 2$ and 6

$$5 + 2 + 6 = 13$$

$6 - 13$ will not work.

Borrow 1 from the thousands to become greater than 13.

$$10 + 6 = 16\ (1000 + 600 = 1600)$$
$$16 - 13 = 3\ (1600 - 1300 = 300)$$

$$\begin{array}{r} 6625 \\ 1597 \\ 284 \\ -\underline{3678} \\ 2386 \end{array}$$

For the thousands. Final step, subtract the remaining from 6.

$$6 - 1 - 3 = 2\ (6000 - 1000 - 3000 = 2000)$$

No borrowing is required on the last step, as the top number is already known to be higher than the numbers below based on previous estimations.

As was fairly certain that the numbers to be subtracted were less than the top number, using this method worked well. Otherwise, simply use one of the original methods or move on to the next chapter to see how this new method is utilised in combination with earlier concepts.

6

The Addition/Subtraction Column Algorithm

When learners are first taught about subtraction, they are told that subtracting a larger number from a smaller number cannot be done or that it is really hard to do. When employing the subtraction column algorithm, the same line of reasoning applies. This column subtraction method was designed in a way that only the larger number could go on top when performing calculations. A larger number refers to the actual number itself, whether or not it has a positive or negative value attached to it.

Upon reaching a more advanced stage, learners will become aware of the presence of two distinct approaches to subtracting a larger number from a smaller number:

1. Disregarding or dropping the subtraction operator and rearranging the numbers into a column subtraction with the larger number on top is by far the most common approach used. The next step here is to subtract the smaller number from the bigger number and then add the negative sign to the answer at the end of the calculations.

2. The utilisation of absolute values as a technique for determining an accurate solution is a rather intricate process that necessitates a considerable amount of expertise in using the concept of absolute values. Due to its complexity, this method will not be discussed any further. However, it is crucial to

acknowledge the existence of this method; and if wanting to learn more about this technique, its accessibility on the internet is readily available.

Both of these methods do work, with the first method being a guess at the end of the calculations. The reason for this is because the negative sign is simply just added to the final answer, and guessing in mathematics is not ideal and the second concept is too difficult for the new learner. Previously in Chapter 3, the addition and subtraction operations were rearranged in order to show how to calculate addition and subtraction in any order; i.e.

$$5 - 2 = 3 \text{ is the same as } -2 + 5 = 3$$

This concept can be used in the addition column algorithm as well as the subtraction column algorithm. In fact, the addition and subtraction algorithms represent a single, unified concept; they function together as a joint process.

When a number is being added, it can be deconstructed, which of course is nothing new or special; it is done so by performing the following method:

$$263 + 481$$

$$200 + 60 + 3 + 400 + 80 + 1$$

This concept can become very useful when teaching addition and allowing learners to perform their own calculations for practice. Similarly, the same concept can be used for deconstructing a number that is being subtracted, as it is exactly the same concept:

$$263 - 481$$

$$200 + 60 + 3 - 400 - 80 - 1$$

It must be stressed that the subtraction of 481 does not at any time make it a negative number, so it should not be changed to adding a negative, which would look like this:

$$263 - 481$$

$$200 + 60 + 3 + {}^{-}400 + {}^{-}80 + {}^{-}1$$

This method does not achieve anything useful to work with at all. The expression is literally subtracting positive numbers: 400, 80, and 1. So when this is applied to the column algorithm, it all becomes very simple to understand and to work on.

Before continuing any further, it is important to know that when subtracting a larger number from a smaller number, borrowing becomes an issue. Using smaller numbers and performing mental calculations means that borrowing is not required. If done correctly, it will always produce a correct answer:

$${}^{-}5 + 14 = 9 \text{ or } 8 - 15 = {}^{-}7$$

The following expression will show working out as it progresses through the subtraction and addition process. The calculations will start from zero to show that numbers are being subtracted from zero; otherwise, it might seem as though the initial addition inexplicably emerges from an unknown source. Exactly as demonstrated in Chapter 3, when rearranging the expression to place the larger value on top, the subtraction operator must be moved with it to maintain the integrity of the calculation:

$$1581 - 7329$$

After rearrangement, the expression looks quite unfamiliar, but the calculations can now be done with ease. Although not really needed, four zeros are provided in the first set of calculations just to show clarity of what the process will look like:

$$0 - 7329 + 1581$$

$$\begin{array}{r} 0000 \\ -\ 7329 \\ \underline{+\ 1581} \\ {}^{-}\ \ 8 \end{array} \quad 0 - 9 + 1 = {}^{-}8. \text{ No need to borrow}$$

$$\begin{array}{r} 0 \\ -\ 7329 \\ \underline{+\ 1581} \\ {}^{-}\ \ 48 \end{array} \quad 0 - 12 + 8 = {}^{-}4. \text{ Borrow 1 from the 3 } (300)$$

$$\begin{array}{r} 0 \\ -\ 7229 \\ \underline{+\ 1581} \\ {}^{-}748 \end{array} \quad 0 - 12 + 5 = {}^{-}7. \text{ Borrow 1 from 7 } (7000)$$

$$\begin{array}{r} 0 \\ -\ 6229 \\ \underline{+\ 1581} \\ {}^{-}5748 \end{array} \quad 0 - 6 + 1 = {}^{-}5. \text{ No need to borrow}$$

This method moves beyond the traditional limitations of column subtraction. Instead of treating the negative sign as a final guess or an afterthought, this approach integrates the negative identity into every column. By intentionally borrowing to maintain sign consistency, the algorithm produces a unified result that is logically grounded from start to finish.

The following expression is a combination of addition and subtraction, generally dealt with by doing one calculation at a time. To start work on the expression, move all of the numbers around, keeping in mind to move the operator with the number directly following it:

$$271 - 63 - 321 + 912 - 391 - 712 + 112$$

Rearranging the expression will show the combined addition and subtraction operators together. Attempting to convert subtraction into the addition of a negative creates significant systemic problems and contradictions within the calculation. Zero is placed at the start of the expression to show that the succeeding number is doing something, which is adding or subtracting. All of this will be shown in the addition/subtraction column algorithm.

$$0 + 912 + 112 + 271 - 712 - 63 - 391 - 321$$

$$\begin{array}{r} 0 \\ +\,912 \\ +\,112 \\ \underline{+\,271} \\ +\,1295 \end{array} \qquad \begin{array}{r} 0 \\ -\,712 \\ -\,\ \ 63 \\ -\,391 \\ \underline{-\,321} \\ -\,1487 \end{array}$$

When showing that all the numbers are being subtracted from zero, generally, only one subtraction operator is placed before the last number. The subtraction operators are there to show clarity as is done with the addition operators. The subtraction operators are not negative signs. Even though the answer of

1487 appears to be negative there is no need to show a negative sign because subtraction and addition are still being done. Every number in both columns are positive numbers.

Following the completion of the addition and subtraction calculations, a final step remains: the two outcomes are compared to determine which holds the greater value. They are then arranged within the addition/subtraction algorithm to produce the final result. After arranging the two numbers to ensure that the higher number is placed first, do the appropriate calculations to acquire a result and gain a final answer. There is no real need to start from zero anymore as long as it is perfectly understood that all calculations do start from zero at some point.

$$\begin{array}{r} -\,1487 \\ \underline{+\,1295} \\ {}^{-}192 \end{array}$$

$$271 - 63 - 321 + 912 - 391 - 712 + 112 = {}^{-}192$$

7

Is the Square Root of a Negative Number a Real Possibility?

In this framework, the term Territory refers to the fixed identity of a number relative to zero. Territorial Integrity is the principle that a calculation remains within its starting identity (either positive or negative) unless it performs a cross-border operation—a movement that passes through zero to enter the opposing Territory.

All the content presented so far has focused on the current functioning of mathematics, particularly at a fundamental level. Moving into a slightly more complicated area of mathematics, it is quite intriguing to observe that while division and multiplication are relatively straightforward operations, there seems to be no simple solution for the square root of negative numbers. When dealing with the square root of a negative number or unit, the use of the imaginary unit, represented by 'i', provides a solution. Thus, the square root of -4 can be represented as $2i$. The standard definition of the imaginary unit is presented as such:

$$i^2 = -1,$$

This equation introduces a level of complexity that exceeds the foundational simplicity established in this framework. As an understanding of this concept is not required at this point, it will not be discussed until a later chapter. The inclusion of an imaginary number in a solution raises the question of why the square roots of negative numbers are not as straightforward as those of positive numbers. This chapter will delve deeply into the enigmatic realm of negatives in order to unravel the answer to a question that has intrigued countless individuals for centuries.

Taking into account the reader has an understanding of square roots; the first aspect to explore is the result of employing negative numbers in conjunction with various concepts. The examination of outcomes resulting from diverse operations involving subtraction and negative numbers presents an intriguing and enlightening prospect. This exploration may potentially unveil solutions or, at the very least, provide a few valuable clues.

Investigation of Negative Numbers Raised to Positive Exponents

A negative multiplied by a negative is positive, and numerous proofs exist to demonstrate this result. In current mathematical practice, standard minus signs are used to represent both the operation of subtraction and the negative identity of a number. This dual use of the symbol often obscures the logic of the calculation. For instance, if negative two is multiplied by negative two, it produces the product of four, which can also be seen in an exponential form:

$$(-2)^2 = -2 \times -2 = 4$$

This seems perfectly logical and quite straightforward but of course this result is not to be confused with $-(2^2)$ which has a different answer and is not part of this topic. While some may perceive negative signs as subtraction operators, the implementation of superscript negative signs will persist in order to ensure comprehension and clarity.

The pattern below shows a common negative number as a base and an increasing positive exponent to analyse and compare how negative numbers behave:

$$({}^{-}3)^2 = {}^{+}9 = {}^{-}3 \times {}^{-}3$$
$$({}^{-}3)^3 = {}^{-}27 = {}^{-}3 \times {}^{-}3 \times {}^{-}3$$
$$({}^{-}3)^4 = {}^{+}81 = {}^{-}3 \times {}^{-}3 \times {}^{-}3 \times {}^{-}3$$
$$({}^{-}3)^5 = {}^{-}243 = {}^{-}3 \times {}^{-}3 \times {}^{-}3 \times {}^{-}3 \times {}^{-}3$$

$$({}^{+}3)^2 = {}^{+}9 = {}^{+}3 \times {}^{+}3$$
$$({}^{+}3)^3 = {}^{+}27 = {}^{+}3 \times {}^{+}3 \times {}^{+}3$$
$$({}^{+}3)^4 = {}^{+}81 = {}^{+}3 \times {}^{+}3 \times {}^{+}3 \times {}^{+}3$$
$$({}^{+}3)^5 = {}^{+}243 = {}^{+}3 \times {}^{+}3 \times {}^{+}3 \times {}^{+}3 \times {}^{+}3$$

The pattern for negative bases continues being positive and then negative for every even and odd number respectively. Even though this pattern is accepted in the world of mathematics, it is far from logical and has no possible way to provide real life examples to explain why this is occurring. When a common positive base is used with increasing positive exponents the same physical numbers appear but they do not follow the same alternating pattern that is seen with negative base results. These results simply stay as positives for all calculations, as they should. Furthermore, providing additional help for learners by way of using addition to see how exponents work becomes quite concerning:

$$\begin{aligned}({}^{+}3)^2 &= {}^{+}9 \\ &= {}^{+}3 \times {}^{+}3 \\ &= 0 + ({}^{+}3) + ({}^{+}3) + ({}^{+}3)\end{aligned}$$

This works well for addition. Because multiplication is repeated addition, the operators must remain. Positive signs within the parentheses represent the value of the multiplicand. Straight repeated addition for a negative squared, results in a negative value; this explains why it is omitted from textbooks and the internet:

$$\begin{aligned}({}^{-}3)^2 &= {}^{+}9 \\ &= {}^{-}3 \times {}^{-}3 \\ &\neq 0 + ({}^{-}3) + ({}^{-}3) + ({}^{-}3)\end{aligned}$$

Even though addition operators should be present because multiplication is repeated addition, it is still unclear if this is correct or not. It does seem that there should be subtraction operators instead, because if they are not present, there is no way of including the other negative sign in the calculations:

$$\begin{aligned}({}^{-}3)^2 &= {}^{+}9 \\ &= {}^{-}3 \times {}^{-}3 \\ &= 0 - ({}^{-}3) - ({}^{-}3) - ({}^{-}3)\end{aligned}$$

Using repeated subtraction to perform the multiplication of two negatives works accurately for even exponents, but fails for odd exponents, as demonstrated below:

$$\begin{aligned}({}^{-}3)^3 &= {}^{-}27 \\ &= {}^{-}3 \times {}^{-}3 \times {}^{-}3 \\ &\neq 0 - ({}^{-}3) - ({}^{-}3) - ({}^{-}3) - ({}^{-}3) - ({}^{-}3) - ({}^{-}3) - ({}^{-}3) \\ &\quad - ({}^{-}3) - ({}^{-}3)\end{aligned}$$

Performing calculations using repeated addition, as shown in the following example of an odd exponent, will yield the correct result:

$$
\begin{aligned}
({}^{-}3)^3 &= {}^{-}27 \\
&= {}^{-}3 \times {}^{-}3 \times {}^{-}3 \\
&= 0 + ({}^{-}3) + ({}^{-}3) + ({}^{-}3) + ({}^{-}3) + ({}^{-}3) + ({}^{-}3) + ({}^{-}3) \\
&\qquad + ({}^{-}3) + ({}^{-}3)
\end{aligned}
$$

This of course is not consistent with the even exponent calculations, and those calculations cannot change just to make something else work. This shows that using negatives with exponents brings forth a serious problem one of which must be addressed by finding out why it is not working.

When working with square roots, calculations once again become strange:

$$x = \sqrt{9}$$

There is only one possible real answer:

$$x = 3$$

But when presented with the inverse of the above:

$$x^2 = 9$$

There will be two possible real solutions:

$$x = 3 \text{ or } x = {}^{-}3$$

This strange notation is completely accepted in the world of mathematics and becomes even stranger when the values change to something as simple as a cube root:

$$x = \sqrt[3]{8}$$

The only possible real answer is:

$$x = 2$$

Once again the inverse is used to compare:

$$x^3 = 8$$

This time the answer is very different with only one possible *real* answer:

$$x = 2$$

But, in this scenario, there are two additional results with complex values involving imaginary components. Due to their abstract structure, they lack a direct physical comparison to real quantities and are omitted here to maintain the focus on tangible results. These examples further highlight the systemic inconsistencies within standard notation.

Continuous Multiplication of Positive and Negative Values

To expose negative inconsistencies, a number is multiplied multiple times by the same constant. The first two sets use a positive two constant, while the third and fourth use a negative two for the calculations:

$$3 \times 2 = 6$$
$$6 \times 2 = 12$$
$$12 \times 2 = 24$$
$$24 \times 2 = 48$$

A positive constant and a positive number do not create any issues for the products.

$$^{-}3 \times 2 = {}^{-}6$$
$$^{-}6 \times 2 = {}^{-}12$$
$$^{-}12 \times 2 = {}^{-}24$$
$$^{-}24 \times 2 = {}^{-}48$$

A positive constant and a negative number to begin with, once again have perfect results.

$$3 \times {}^{-}2 = {}^{-}6$$
$$^{-}6 \times {}^{-}2 = 12$$
$$12 \times {}^{-}2 = {}^{-}24$$
$$^{-}24 \times {}^{-}2 = 48$$

These calculations clearly show that it is failing to keep a constant pattern.

$$^{-}3 \times {}^{-}2 = 6$$
$$6 \times {}^{-}2 = {}^{-}12$$
$$^{-}12 \times {}^{-}2 = 24$$
$$24 \times {}^{-}2 = {}^{-}48$$

A negative constant and a negative number to begin with have the same confusing patterns seen previously. It is clear that these calculations are not working the same way every time, but there is no clue here as to why. Based on the information provided so far, the standard way of performing basic calculations, simply does not explain why these inconsistencies are occurring.

Negative Logarithms

This is only brief look at logarithms, simply because this book is not going to explain how logs work. The focus will be on negative numbers to see how they behave when used in logarithm calculations. Just like the above information for exponents, but performed inversely, the calculations for some negative numbers work perfectly while other negative numbers do not:

$$log_{^-3} {}^-27 = 3$$

This is log $^-27$ base $^-3$ which equals 3. Even though negatives are considered to have no real solution for logarithms, when inverting back to an exponentiation equation it certainly shows that it does work:

$$(^-3)^3 = {}^-27$$

When using an even number for the base, the calculations for negatives are now incorrect:

$$log_{^-2} {}^-4 \neq 2$$

If many more examples were displayed, it would show that all the odd base numbers work while the even base numbers display the wrong answer.

Negative Division

Upon looking at how to divide a larger negative number, there appears to be little information on how to approach this. Keep in mind that division is the inverse of multiplication which means that division is repeated subtraction. Most articles found (none credible or published) advise to drop the signs, work out the division, and then add the necessary positive or negative sign at the end of the calculations. This approach is just guessing and not a mathematical way of working out a problem.

Below is an example of how long division with a negative would look:

$$^{-}126 \div {^{+}2}$$

$$\begin{array}{rl}
{^{-}63} & \\
{^{+}2}\overline{)\,{^{-}126}} & {^{-}12} \div {^{+}2} = {^{-}6} \\
\underline{-\,{^{-}12}} & {^{+}2} \times {^{-}6} = {^{-}12} \\
{^{-}06} & \\
\underline{-\,{^{-}6}} & {^{-}6} \div {^{+}2} = {^{-}3} \\
0 & {^{+}2} \times {^{-}3} = {^{-}6}
\end{array}$$

This approach does work and if the signs were changed over from a positive dividend to a negative and a negative divisor to a positive the calculations will still work.

$$^{+}126 \div {^{-}2}$$

However, for this example if division is used as repeated subtraction, the results would see the calculations head towards negative infinity: $^{-}126 - {^{+}2} - {^{+}2}$...
Division with two negatives can be seen as follows:

$$^{-}126 \div {^{-}2}$$

$$\begin{array}{rl}
{^{+}6} & \\
{^{-}2}\overline{)\,{^{-}126}} & {^{-}12} \div {^{-}2} = {^{+}6} \\
\underline{-\,{^{+}12}} & {^{-}2} \times {^{-}6} = {^{+}12} \\
{^{-}24} &
\end{array}$$

There is no point going further with the calculations simply because the partial subtraction results are moving towards negative infinity. The result should be heading towards zero. The calculations are taking into account that division is repeated subtraction or with long division it is subtraction that uses a chunking method, which is still subtraction no matter how it is interpreted. As shown, a negative divided by a negative does not work, and one of the bigger issues is that division only subtracts towards zero and never passes from one side of zero on the number line to the opposite side. Division just does not work like that in any way possible. It would be like saying that $^{+}24 \div {}^{+}6 = {}^{-}4$, it does not work.

The Search for Consistency

Initial investigations into why these concepts fail have yet to yield a consistent explanation for the inconsistencies found in multiplication and division. It is difficult to pinpoint exactly why the additive and subtractive processes within these operations stop making sense. In reality, the laws of multiplication and division simply follow in the footsteps of addition and subtraction. It is more accurate to say that the combination of symbols in the laws for addition and subtraction forced multiplication and division to adopt identical rules for the sake of consistency. This implies that any fundamental error stems from the very foundation of addition and subtraction; therefore, the investigation begins by delving deeply into that world to gain fresh insights. Because the concept of positive is understood without doubt, it will remain a constant anchor for comparison.

Start of the investigation

Addition and Subtraction with Positive and Negative Numbers

The following are what is known to be fact, either from axioms, proofs, logic as well as general common knowledge. Every effort has been undertaken to ensure the utmost precision by avoiding any ambiguous terminology or expressions.

- On the positive side of the number line, addition moves to the right and subtraction moves to the left.
- On the negative side of the number line, addition moves to the left and subtraction moves to the right.
- Actual physical numbers increase to the right on the number line in a positive direction.
- Actual physical numbers increase to the left on the number line in a negative direction.
- There are four operators, addition, subtraction, multiplication and division.
- Addition and subtraction are inverses of each other. Multiplication and division are also inverses of each other.
- Multiplication is repeated addition and division is repeated subtraction.
- All numbers are positive or negative, all numbers exist on a number line; therefore numbers on a number line are either positive or negative.
- Zero is considered neither negative nor positive, but when being positioned as a place holder as the last digit of a number, that whole number is considered positive or negative.
- Zero is considered neither odd nor even but when being positioned as a place holder as the last digit of a number, that whole number is considered even. Zero also lies between two odd numbers but is still considered neither odd nor even.
- Addition, subtraction, multiplication and division are always operators and never possess a positive or negative value.
- All numbers, except zero, are positive or negative and are never used as operators.

- Numbers and the concept of positive and negative are separate ideas, which conjoin to form a value of a number in space and placement within the world of mathematics i.e. something can be negative without knowing the value of it and vice versa.

Concepts within mathematics need to have consistency and must only work in the way they were intended. This means that abstract functionality cannot be imposed on something simply because someone needs it to function in a specific way. If mathematics is forced to work like that, then it will eventually fail, even with the smallest of concepts associated with it. If a mathematical concept works in a very specific manner, it must also work inversely or in an opposite way. For example, if the concept of counting up is possible then the concept of counting down too must be possible and both will be either an inverse or an opposite concept of each other at all times. The reason why inverses and opposites exist in mathematics is because of the way that mathematics has been created by humans. A simple example of this is the invention of positive values and how they appear on the number line, which is seen as moving to the right of zero. When negatives were introduced on the other side of zero, it forced the concept of opposite to occur. This in turn has created a foundation of number recognition of opposites for positives and negatives that cannot be undone. For all concepts that occur in mathematics an opposite or an inverse will be present which is important to know as these very important ideas will be utilised in later explanations of concepts. The definitions of inverses and opposites will be explored further in this chapter. The very first concept to examine is; what is happening when two numbers are combined together. Axioms, corollaries, and postulates exist in mathematics to help create a definition of addition, but do not need to be understood here, just understanding that they exist is sufficient. Later in this chapter the addition definition will be discussed in a simple to understand manner.

One of the greatest teaching aids that are used in mathematics is the number line and for this reason it will be used continuously and will help to show consistency throughout this chapter. The number line will be discussed in greater detail for more advanced calculations towards the end of Chapter 8.

The Augend, Addend, Minuend and Subtrahend

In any mathematical expression, numbers occupy specific roles based on the operation being performed. These are defined as:

- **Augend:** The starting quantity to be increased.
- **Addend:** The quantity being added to the augend.
- **Minuend:** The starting quantity from which another is subtracted.
- **Subtrahend:** The quantity to be subtracted from the minuend.

Technical Note: While modern curricula often generalise addition terms as addends and subtraction terms as parts, this book maintains these traditional distinctions. This precision is necessary to identify the exact starting point—the logical anchor—of every calculation.

The structural significance of these four concepts is of great importance, as each one serves as an exact instruction for the expression to be carried out with the utmost precision. When retrieving information from a mathematical story, each value or variable is placed into the expression, which is precisely when it is assigned its specific functional name. They represent the actual units or objects being calculated to determine a total quantity. Furthermore, these augends, addends, minuends, and subtrahends are inherently either positive or negative, with the sole exception of zero. When a calculation involves both addition and subtraction, the final answer is no longer confined to the labels of sum or difference. It becomes the 'Total Quantity'—the final state of the information after every augend, addend, minuend, and subtrahend has played its part in the mathematical story.

The Addition and Subtraction Operators

The addition operator imposes two conditions on the expression. First, it instructs the calculations to be combined to produce a greater amount of a given quantity. Second, it dictates the direction in which the calculations progress on a number line—moving towards either positive or negative infinity. When the addition operator instructs the calculations to face in the direction towards infinity, the amounts indicated by the augend and the following addends move that required distance. Conversely, the subtraction operator reverses this spatial progression, instructing the calculations to move in the exact opposite direction along the number line.

Every operator within an expression must always have a value before it and a value after it. By logical reasoning and common sense, two operators never occur consecutively. If two symbols appear together, the symbol attached to the number following the operator is always a positive or negative sign, rather than a second operator.

The Positive and Negative Signs When Adding and Subtracting

Positive numbers exist only on the positive side of the number line, and negative numbers exist only on the negative side. This is an absolute fact. These signs perform one function: they state the identity of the object, which indicates the side of the number line where the calculation takes place. This is represented as a superscript sign positioned after the operator and before the number.

Typically, determining the side of the number line is done before applying the operators but this is not essential. Standard sign rules are often used as shortcuts to speed up the working process. While these shortcuts appear to be a convenient way to deal with addition and subtraction followed by a negative or positive sign, the narratives accompanying these shortcuts are logically incorrect. Two factors contribute to this inaccuracy. First, addition and subtraction symbols are operators; they are not signs belonging to the preceding numbers. Although this is well-known, they are still occasionally mislabelled as such. Second, the names given to these shortcuts suggest that the signs play a role other than simply identifying the number. For instance, it is commonly taught that two identical signs make a positive and two opposite signs make a negative. While beneficial for speed, these statements obscure the underlying logic, particularly when multiple similar symbols appear in the same expression. These signs create nothing new; they simply indicate the direction of the calculation. Commonly described as plus and minus, addition and subtraction, or positive and negative, these symbols are often used interchangeably. However, they are rarely utilised as superscripts. The following are not absolute, but represent a common set of examples.

Directional Movements of Plus and Minus Signs

- Plus sign and a plus sign: Equals addition and moves towards positive infinity.
  $+(+) = +$ (e.g. $3 + (+2) = 3 + 2$)
- Plus sign and a minus sign: Equals subtraction and moves towards negative infinity.
  $+(-) = -$ (e.g. $-3 + (-2) = -3 - 2$)
- Minus sign and a plus sign: Equals subtraction and moves towards negative infinity.
  $-(+) = -$ (e.g. $3 - (+2) = 3 - 2$)
- Minus sign and a minus sign: Equals addition and moves towards positive infinity.
  $-(-) = +$ (e.g. $-3 - (-2) = -3 + 2$)

Establishing the Base Model for Addition and Subtraction

The following steps are built upon the foundational truths established from the start of this investigation. This framework serves as the base model for every example encountered throughout the remainder of this book.

To ensure absolute clarity, each number within the expression is identified by its specific functional role before the calculation begins.

For the first example:

$$0 + {}^{+}3 + {}^{+}4$$

The roles are defined as follows:

- **Zero $(\mathbf{0})$:** The Augend (The fixed starting point).
- **Positive Three $({}^{+}\mathbf{3})$:** The first Addend.
- **Positive Four $({}^{+}\mathbf{4})$:** The second Addend.

The calculation is then performed one step at a time using the directions from the base model. Because zero is the augend, it is the established anchor and requires no calculation. The first operation is applied to the first addend $({}^{+}\mathbf{3})$, followed by the second addend $({}^{+}\mathbf{4})$, resulting in the final total.

The following model describes the addition of $({}^{+}\mathbf{3})$.
The sign of the addend, which is positive, states which side of the number line that the calculations will take place, which is the positive side. The addition operator states which way to face on the number line once the correct side of the number line has been determined, which is towards positive infinity. The addend, $^{+}3$, confirms how far to move on the number line which is 3 positive units, 1 unit at a time, from the augend, 0.

The following model describes the addition of $({}^{+}\mathbf{4})$.
The sign of the addend, which is positive, states which side of the number line that the calculations will take place, which is the positive side. The addition operator states which way to face on the number line once the correct side of the number line has been determined, which is towards positive infinity. The addend, $^{+}4$, confirms how far to move on the number line which is 4 positive units, 1 unit at a time, starting from the position established in the previous model: $\mathbf{0} + {}^{+}\mathbf{3} = {}^{+}\mathbf{3}$.

The steps taken are absolute but the order of those steps may differ, for example addition may be determined which way to face on the number before the positive sign is determined. This difference has no effect on the process, the sign could be negative and if determining that addition is going to happen first then this means that the calculations will be facing towards infinity in a positive or negative direction. But the addend cannot move until the initial two actions are taken first.

The same model will be preserved and utilised for the rest of the addition/subtraction part of the chapter as consistency is paramount. This will allow all the combinations of the signed numbers with both operators to be true and correct at all times. The correctness of the base model will then be able to be extended onto multiplication and division as both are repeated addition and repeated subtraction respectively. The reason this concept can be extended to multiplication and division is that the calculations utilise the exact same process; however, this process differs significantly from traditional interpretations, as will be explored shortly.

All of this looks so simple, but when being consistent and never wavering from these basic facts, it will help to see and understand more accurately as to what is happening with other calculations. Having this knowledge will help to understand how subtraction and addition, along with positive and negative numbers work together. Being consistent will help to unravel what is going on in the mathematics of today.

The next example, like the first, is fairly simple, but a little less intuitive as it is used somewhat less often than addition. The mathematical story may look something like "I have five apples and I give two of them away to a friend". This is how a subtraction expression may be seen and written when being taught.

$$5 - 2$$

This could be read as five minus two.

Again the expression is rewritten with all the missing but necessary information.

$$0 + {}^{+}5 - {}^{+}2$$

This example could now be read as zero plus positive five, minus positive two.

As with the base model example for addition, first comes the mathematical story then the expression or the entire equation. The calculations begin as before, by following the instructions provided by the operators and the addend signs. In the problem above, five apples have been obtained in some way which is addition. Addition is the only possibility because the apples were obtained after previously having zero apples. Again these five apples are the same as before which is positive. Two of them were given to a friend, so there are two less, which is counting back and becoming less than five. This of course is subtraction as two apples are being taken away from the original five apples.

Implicit knowledge dictates that subtraction is the inverse of addition; as such, the inverse action undoes the initial action, which is a process fundamentally different from the definition of an opposite. Even though only a partial amount has been subtracted it is still the inverse of addition. If this was checked on a number line it could be seen as the very basic movements of subtracting two from five. More importantly it can be seen that the calculations are still on the positive side of the number line. The answer is still positive, and the amount that was subtracted was positive. When revisiting the base model of addition it is plain to see that the same values are still being calculated, that being positive, so nothing has changed.

Negative values have not appeared which there should not be and this is keeping the observations consistent. It must be noted that while the subtraction operator changes the direction of the calculations to move towards zero or becoming less, the end result shows that it does not in any way interfere with or change the positiveness of the two apples being subtracted, they remain as positive values at all times. Changing the subtraction to adding a negative would mean that two negative apples would need to be added, which of course is impossible for this situation. On the number line it is very clear that literally there are positive numbers being subtracted, again this is very important to understand as being consistent is essential. The other important concept is that the positive signs are not directing the calculations to go in a different direction. This is mainly referring to the subtraction of the positive subtrahend as the calculations move in what is perceived to be a negative direction towards zero. Omitting the positive signs again will work for more advanced users and once again the answer is a positive and should not be written as an operator symbol.
For the next example:

$$0 + {}^{+}5 - {}^{+}2$$

The roles are defined as follows:

- **Zero $(\mathbf{0})$:** The Augend (The fixed starting point).
- **Positive Five $({}^{+}\mathbf{5})$:** The Addend.
- **Positive Two $({}^{+}\mathbf{2})$:** The Subtrahend.

There is no need to model $({}^{+}\mathbf{5})$ as it is the same process as seen in the previous model demonstration.

The following model describes the subtraction of $(^{+}2)$.
The sign of the subtrahend, which is positive, states which side of the number line that the calculations will take place, which is the positive side. The subtraction operator states which way to face on the number line once the correct side of the number line has been determined, which is towards zero. The subtrahend, $^{+}2$, confirms how far to move on the number line which is 2 positive units, 1 unit at a time, starting from the addend: $0 + {}^{+}5 = {}^{+}5$.

The next concept involves calculating on the negative side of the number line. It is essential to remain consistent with what has been observed and known to be true so far by utilising the base model. Consider the following mathematical story: "If it is negative eight degrees Celsius in the freezer and the temperature increases by three degrees, what is the temperature now?"

$$^{-}8 + 3$$

Adding the extra information will show the expression with greater clarity:

$$0 + {}^{-}8 + {}^{+}3$$

The first expression looks the same as it always has, but there is a problem with a positive three being added to a negative. Adding positives on the negative side of the number line is inconsistent, as there are no positive numbers there. Writing the expression correctly requires remaining consistent and using what is already known to be true. The actual numbers on the negative side of the number line increase to the left, towards negative infinity, which is the addition of negative numbers. Actual numbers on the negative side of the number line decrease towards zero, which is the subtraction of negative numbers. Therefore, only the subtraction of negative numbers towards zero is possible, as positive numbers do not exist there.

This concept is not an alternative to make it look different or better; it is the actuality of mathematics. Although once a learner becomes adept with their knowledge and proficiency in mathematics, shortcuts may be used—which, of course, is fine—this is not a lesson on how to learn or teach those shortcuts. Showing what is really happening and being consistent is the only way to ensure that all calculations are perfectly clear, precise and very well understood.

The expression can now be completed correctly by changing the addition of a positive to the subtraction of a negative. The original mathematical story—"If it is negative eight degrees Celsius in the freezer and the temperature is increased by three degrees, what is the temperature now?"—now has greater meaning. After adding negative eight, the temperature still needs to be raised by three degrees.

To show how the expression is changed it will be written as such:

$$0 + {}^{-}8 - {}^{-}3$$

The roles are defined as follows:

- **Zero $(\mathbf{0})$:** The Augend (The fixed starting point).
- **Negative Eight $({}^{-}\mathbf{8})$:** The Addend.
- **Negative Three $({}^{-}\mathbf{3})$:** The Subtrahend.

The following model describes the addition of $({}^{-}\mathbf{8})$.

The sign of the addend, which is negative, states which side of the number line that the calculations will take place, which is the negative side. The addition operator states which way to

face on the number line once the correct side of the number line has been determined, which is towards negative infinity. The addend, $^{-}8$, confirms how far to move on the number line which is 8 negative units, 1 unit at a time, from the augend, 0.

The following model describes the subtraction of $(^{-}3)$.
The sign of the subtrahend, which is negative, states which side of the number line that the calculations will take place, which is the negative side. The subtraction operator states which way to face on the number line once the correct side of the number line has been determined, which is towards zero. The subtrahend, $^{-}3$, confirms how far to move on the number line which is 3 negative units, 1 unit at a time, starting from the position established in the previous model: $0 + {}^{-}8 = {}^{-}8$.

Mathematics has allowed for turning a positive number into a negative number and as discussed earlier it is not possible to do it as negatives do not exist on the positive side of the number line. Below is how it will look if attempting to use the base model directions:

$$6 - 2$$

$$6 + (-2)$$

$$0 + {}^{+}6 + {}^{-}2$$

The roles are defined as follows:

- **Zero** (0)**:** The Augend (The fixed starting point).
- **Positive Six** $(^{+}6)$**:** The Addend.
- **Negative Two** $(^{-}2)$**:** The Addend.

The following model describes the addition of $(^{+}6)$.
The sign of the addend, which is positive, states which side of the number line that the calculations will take place, which is the positive side. The addition operator states which way to face on the number line once the correct side of the number line has been determined, which is towards positive infinity. The addend, $^{+}6$, confirms how far to move on the number line which is 6 positive units, 1 unit at a time, from the augend, 0.

The following model describes the addition of $(^{-}2)$.
The sign of the addend, which is negative, states which side of the number line that the calculations will take place, which is the negative side. The current calculations of the positive six have just taken place on the positive side of the number line. Now the calculations must jump to the negative side of the number line to add negative two. This of course is not possible and the calculations must cease because of this error.

These models provide a definitive and singular way to represent addition and subtraction correctly. Because they are grounded in mathematical actuality rather than mere shortcuts, they function as the necessary blueprint for the chapters ahead. This established framework will be used to demonstrate how multiplication and division operate with the same perfect clarity.

Calculating Addition and Subtraction Through Zero

The transition across zero on the number line represents a critical juncture in mathematical logic. Standard teaching methods often overlook the mechanics of this crossing, leaving a conceptual gap that becomes particularly problematic when moving into multiplication and division. In those higher-level operations, values frequently shift from the negative side to the positive—and vice-versa—without a clear explanation of the underlying process.

By defining the precise movement required to pass through zero using addition and subtraction, the structural consistency of the number line is maintained. This section provides the necessary clarity for these transitions, ensuring that when multiplication and division are introduced, the movement across zero is already understood as a logical certainty rather than an unexplained leap.

The challenge of explaining an expression that transitions from one side of zero to the opposite side remains one of the most significant hurdles in mathematical instruction. Consider the following example. Including all the necessary information, this expression may look strange initially. Despite being a departure from previous steps, it is a vital inclusion that remains quite simple to follow:

$$0 + {}^{+}4 - {}^{+}7$$

While this looks straightforward, problems are inevitable when a smaller number subtracts a larger number. The issue of passing through zero means that, at some stage, the subtraction of positive numbers is going to occur on the negative side of the number line. To prevent this problem from occurring, the calculation of subtracting positive seven can be deconstructed to show exactly what must happen to allow the calculations to work properly. Working the problem out according to a number line, looks like this:

$$0 + {}^{+}4 - {}^{+}4 + {}^{-}3$$

An example of passing through zero from the opposite direction—from negative to positive—shows a similar problem:

$$0 + {}^{-}5 - {}^{-}9$$

When changing the calculations to match the transition through zero, it extends to this:

$$0 + {}^{-}5 - {}^{-}5 + {}^{+}4$$

A mathematical story example could be: The temperature was zero and turned down to negative five degrees. What would the temperature be if it was then turned up by nine degrees? The temperature is ${}^{-}5$. To get to 0 degrees, the temperature now needs to go up by 5 degrees. This is subtraction, not addition; it is literally the removal of cold (taking negative degrees away from the existing negative degrees). The next step is for the temperature to rise further on the positive side of zero to ${}^{+}4$ degrees, which is where the ${}^{+}4$ is added. These appear to be two separate events, but that is only for visualising the process of turning a dial and moving through zero in a single, fluid motion.

This shows that combining a subtraction operator and a negative sign can use the shortcut of turning it into a positive, which moves the calculation in a positive direction. While this shortcut can certainly be used, the information here is for discovering how these calculations really work, rather than just teaching the shortcuts.

It must also be understood that this cannot be done while on the positive side of the number line. For example,

$${}^{+}7 - {}^{-}3$$

This cannot be done—not because it is mathematically impossible in the 'Old Maths,' but because it cannot be done in the physical world. There is no mathematical story that supports subtracting negative three of something from a positive quantity on the positive side of the number line. A negative three of something cannot exist in that territory. It would be like saying, "I have seven apples and I am trying to take away three oranges." Since the oranges are not there to begin with, it is simply impossible to take them away.

From what has been seen in some of the previous examples, it is very clear that adding and subtracting positive and negative numbers is very literal in the real world. The same combinations previously examined are presented below; however, the transition highlights the precise language that must be applied to each instance.

Adding a positive equals a positive result and can only be added from the positive side of zero, calculations move towards positive infinity:

$$+^{+} = {}^{+}$$

Subtracting a negative equals a negative result and can only be subtracted from the negative side of zero, calculations move towards zero but never through zero.

$$-^{-} = {}^{-}$$

Adding a negative equals a negative result and can only be added from the negative side of zero, calculations move towards negative infinity.

$$+^{-} = {}^{-}$$

Subtracting a positive equals a positive result and can only be subtracted from the positive side of zero, calculations move towards zero but never through zero.

$$-^{+} = {}^{+}$$

As the focus shifts to multiplication and division, it remains essential to remember that operations involving negative values cannot originate from the positive side of the number line. While 'Old Maths' utilises shortcuts—such as using absolute values to measure distance—the purpose of this work is to expose the mechanical actuality of the calculation. By removing these shortcuts, the missing information is restored, allowing the process to be seen exactly as it functions in the physical world.

Multiplication and Division with Positive and Negative Numbers

What is Multiplication?

While multiplication is traditionally described as adding a number to itself, that phrasing is open to interpretations that can lead to structural ambiguity. To ensure a higher standard of clarity, this book defines multiplication as **the process of adding the same value to zero a specified number of times.** By establishing zero as the consistent starting point, the uncertainty found in standard repeated addition is removed, providing a stable logical anchor for every calculation. Crucially, this definition encompasses both positive and negative values; whether the units being added are positive or negative, the process remains a mechanical movement from zero into the respective territory of the number line. Whether multiplication is truly repeated addition and whether it should be taught as such is a debate that extends far beyond the contents of this book. Despite these academic concerns, the concept of repeated addition will be utilised throughout this chapter. For new learners, it is the standard introduction; for advanced users, multiplication is often performed instinctively without questioning its foundational nature. When learners are first introduced to multiplication, basic concrete models are used to convey information clearly.

The very word Multiplication reveals its structural truth: it is the Multi-Application of addition. It is not a separate species of mathematics, but simply the process of addition being applied multiple times to a specific group environment.

A number multiplied by another number consists of three basic parts:

1. **The Multiplicand:** The number being multiplied to obtain a result, and also represents something specific and real, such as a physical object or a unit.
2. **The Multiplier:** The value that dictates how many times the multiplicand will be processed.
3. **The Multiplication Operator:** The symbol that initiates the multiplication process, commanding the multiplier to act upon the multiplicand.

Key concepts utilised within this chapter include positive, negative, addition, and the final result, known as the product. Understanding the distinctions between these concepts is vital to eliminate ambiguity; therefore, each will maintain its own unique and consistent meaning.

Mathematics is more than a series of problems to be solved without thought; it possesses inherent depth and context. Just as with addition, every multiplication problem originates from a real-life situation. Translating these situations into words allows the underlying logic to become visible. While learners are rarely asked to write these mathematical stories to provide context from a number sentence or an expression, doing so is an essential skill for mastering advanced mathematics.

In standard practice, the multiplicand and the multiplier are often treated as interchangeable factors simply to reach a product. While this shortcut works for finding a result, the right way requires the constant identification of each terms specific role. This discipline ensures operational consistency and allows for the capture of the exact moment a logical change occurs. There are two distinct ways to interpret multiplication when interchanging these numbers:

Multiplier First $(\mathbf{3 \times 4})$**:** Three groups with four in each group. This is useful for basic understanding but can be limited in application.

Multiplicand First $(\mathbf{3 \times 4})$: Three increased four times. This provides greater flexibility for scaling and notation. It clearly identifies three as the object being multiplied by four.

In this chapter, the multiplicand will always be written first to maintain consistency and eliminate ambiguity. Peano's axioms for multiplication derive directly from his axioms for addition. Therefore, the discoveries made regarding addition and subtraction will be utilised here. While this investigation does not aim to prove or disprove repeated addition, that framework is utilised as it is the primary method through which learners engage with the concept (see Chapter 10 for further discussion on repeated addition including repeated subtraction). It is a certainty that a positive multiplicand with a positive multiplier produces a positive product and moves in a positive direction. While multiplication is an advanced adaptation of addition, the underlying addition often happens invisibly in the background.

Consider this story: "There are four baskets with three apples in each; how many apples are there in total?" Writing an expression for this is simple and logical:

$$3 \times 4$$

To make the logic explicit:

$$^{+}3 \times {}^{+}4$$

In addition and subtraction, the process always begins at zero. While mental arithmetic often skips this step, the logical process still relies on that starting point. Furthermore, it must be recognised that switching numbers, 4×3 to 3×4, changes the entire mathematical story. The multiplicand becomes the multiplier and vice versa. While this is often done to aid memory recall, it can disguise underlying logical problems especially when moving beyond positive numbers. To visualise this on a number line, the most effective method is segmented jumps. The multiplier indicates the number of jumps, while the multiplicand defines the distance of each jump, typically tracked with labels above the line to show progress.

The Multiplicand, a Little Extra But Important Information

The multiplicand is functionally identical to the augends and addends of addition; it is the core focus of the calculation. The primary objective is to determine the final state of the multiplicand, not the multiplier. Though frequently ignored, the multiplicand and multiplier are more than just factors. If a factor is a constituent part of something, the multiplicand—as the primary identity and substance—is the only true factor, while the multiplier acts as the agent of its change.

The Positive and Negative Signs of the Multiplicand

The signs of the multiplicand provide the exact same information as the signs of the augends, addends, subtrahends, and minuends. They are, at all times, either positive or negative and remain the primary focus of the number sentence. In other words, the sign informs the reader that the multiplicand, the very substance of the calculation, exists either in the positive or negative territory of the number line.

The Multiplier and the Signs of the Multiplier with Greater Clarity

The multiplier performs the same function as the addition operator, but with significantly more command. It magnifies the multiplicand through repeated addition to form a greater total. For a new learner, this is often visualised as groups, baskets, boxes, or even non-physical entities. These groups are real-world interpretations of the multiplier. While a group can be empty (zero or null), it is essential to realise that the multiplier has the ability to *hold* objects but is never *part* of the object count itself. On closer inspection of the number line, the jumps of the multiplier and the count of the multiplicand increase at different rates. For instance, when multiplying three by two, the first group has aa positional step of one (the 1st group), yet it represents three units on the number line. The second group possesses another three units giving a total of six units for both groups. There are two distinct concepts of counting happening simultaneously:

1. **The Tallying of Multiplicands:** A 1: 1 count that corresponds directly to the numbers on the number line.

2. **The Ordinal Count of the Multiplier:** A count of how many groups or how many jumps, which does not coincide with the number line values.

Because these jumps are ordinal (1st, 2nd, 3rd), multipliers are not factors of the initial entities being calculated; they are counts of the process. This is the reality of multiplication. Just as a number is assumed to be positive if it lacks a sign, a basket or group is assumed to exist in a positive environment unless stated otherwise. If a group exists in a negative environment, the mathematical story must describe this clearly so it can be transferred to the expression as a negative multiplier. These environments are subjective—described as moving backward, rotated, under, or out. Defining the environment ensures the multiplier remains in that state until the story changes. Crucially, the multiplier cannot change the state of the multiplicand. A multiplier of $^{-}2$ indicates two groups in a negative space, not the value of $^{-}2$ on a number line, they are quite different.

Now that the multiplier, multiplicand, and their signs are understood, the mechanics of how they truly interact can be revealed. In this model, multiplication is always repeated addition, while division is repeated subtraction. A negative multiplicand is multiplied using repeated addition on the negative side of zero, calculating towards negative infinity. For multiplication to remain consistent, the actions performed by the multiplier must be applied identically to every succeeding calculation.

Multiplication and Division: The Base Model Structure

Multiplication for positive numbers means more, so having more positive numbers means they move towards positive infinity and never towards zero. The numbers on the positive side of the number line literally become greater. The same can be said for negative numbers as well. If multiplication is more, then the actual numbers on the negative side of the number line are becoming greater while moving towards negative infinity, which is exactly the same as addition. Also, keep in mind that a jump refers to the movement on the number line made by the multiplier and when performing division, the divisor.

Every jump represents a calculated commitment, dictated by the sign of the multiplicand and the intent of the multiplier. Unlike 'Old Maths,' where multiplication feels like teleporting, the Base Model treats each jump as a physical transit. This ensures the state of the units—positive assets or negative debts—is preserved. The multiplier does not simply arrive at a product; it systematically constructs it unit by unit, maintaining a perfect tally from the origin.

For Multiplication:

The sign of the multiplicand states which side of the number line that the calculations will take place, which is the same as the sign of the addend. The multiplication operator states which way to face on the predetermined side of the number line. This is towards positive or negative infinity, which is the same as the addition operator and never faces towards zero. The multiplicand asserts how far to move on the number line per jump, which is the same as the addend. The multiplier states how many times the multiplicand adds repeatedly on the number line. The sign of the multiplier states what environment the calculations are taking place.

For Division:
The sign of the dividend states which side of the number line that the calculations will take place, which is the same as the sign of the subtrahend. The division operator states which way to face on the predetermined side of the number line. This is towards zero, coming from either positive or negative infinity but never towards either infinity, which is the same as the subtraction operator. The dividend asserts how far to move on the number line per jump, which is the same as the subtrahend. The divisor states how many times the dividend subtracts repeatedly on the number line. The sign of the divisor states what environment the calculations are taking place.

The above directions of how to perform multiplication and division have been directly obtained from the base model of addition and subtraction. The multiplication base model will be used for the following examples. The base model for division does not need to be shown as it is an exact inverse of the multiplication model.

Example 1: Positive Foundations
The first example that is going to be used is:

$$2 \times 3$$

Earlier it was discussed that both the multiplicand and the multiplier are positive numbers, so when the extra information is added, it will look like this:

$$^{+}2 \times {}^{+}3 = {}^{?}?$$

This can be read in different ways: three groups of two in each group, two multiplied by three, or even two increased three times. Because the multiplicand is written first, to keep being consistent for the rest of this topic, the way in which the mathematical story will be expressed is positive two $(^{+}2)$ multiplied by positive three $(^{+}3)$. It is not the correct way or the only way; it could have been said other ways. The word positive will be used continuously simply because the word negative will be used extensively; so this will help keep everything neat and tidy, and balanced with clarity.

As with addition, it directed the two numbers to be combined together which moved in a positive direction. The examples and their instructions below have not been written in a way to make them work in some specified way; moreover, they have been written in a way that shows exactly how the calculations are really working. Great care has been taken for this information to be written in a manner that is very precise and non-misleading. The information has also been written in an easy-to-understand fashion, not because of wanting to make the information look better but because the way in which multiplication really works. In doing so, it makes the following directions very natural and virtually writes itself.

For the first application of two positive numbers:

$$\mathbf{^{+}2 \times {}^{+}3 = {}^{?}?}$$

Use the following multiplication model to calculate a product and to find out if the sign of that product is going to be positive or negative.

The multiplicand is $\mathbf{^{+}2}$ and the multiplier is $\mathbf{^{+}3}$.

The sign of the multiplicand states which side of the number line that the calculations will take place, which is the positive

side. The multiplication operator states which way to face on the predetermined side of the number line, which is towards positive infinity. The multiplicand states how far to move on the number line per jump, and this is 2 positive units per jump. The multiplier states how many times the multiplicand moves repeatedly on the number line, which is 3 jumps. The sign of the multiplier states what environment the calculations are taking place, which is in a positive environment.

Hence: $^{+}\mathbf{2} \times {}^{+}\mathbf{3} = {}^{+}\mathbf{6}$

Example 2 and 3: The Commutative Assumption

Having a look at the next two number sentences together, it can be seen that they are quite similar but not exactly the same. Yet they yield the exact same result, which is a negative value.

$$^{-}2 \times {}^{+}3 = {}^{-}6 \qquad {}^{+}2 \times {}^{-}3 = {}^{-}6$$

In his book, *Elements of Algebra*, Leonhard Euler briefly explains how a negative multiplied by a positive can be a negative. Unfortunately, he gives no explanation or proof as to why the commutative property of a positive multiplied by a negative is negative as well, and all he states is, "which is the same thing" (Euler, 1770). Of course, Euler did not need to make a proof as commutation was accepted through axiomatic rules as it is today, and there is also a modern formal proof for this. This idea of using axioms for negatives is precarious, as the logic fails (without resorting to adding a negative) when these axioms are applied.

This is where the process of performing one operation a certain way contradicts the commutative way it is done. Through rigorous testing and manipulation of numbers, signs and operators over a very long period of time, one exact same process could not be found that fits both number sentences to make them equal a negative. There appears to be no proof for why and how both number sentences result with the same answer, except that the laws of distribution can show this. One other idea of performing these two calculations is to say that the negative sign flips the direction of the calculations, which is not very mathematical at all. The most common way to rationalise the two number sentences equalling a negative is done simply through commutation, and this is just an assumption that is formed from assumed logical reasoning from other calculations; again, this is not a proof.

To achieve both number sentences resulting in a negative, it is accepted to use the negative number as the multiplier, to make the first calculations or jumps on the number line to the left in a negative direction. The second number sentence uses the negative as the multiplicand and to count single steps on the number line, negative two at a time. This of course sends it in a negative direction of three sets. These two sentences are of course treated differently, because it is the only way they can work. When teaching these two concepts to learners, very little instruction or explanations are used, because in traditional instruction, these conceptual contradictions are rarely addressed, as a lack of a cohesive model makes them difficult to reconcile. Consequently, these hurdles are often bypassed entirely, leaving the learner to rely on memorised rules rather than a grounded understanding of how the calculations move.

Use the following multiplication model for the second application:

$$^{+}\mathbf{2} \times {}^{-}\mathbf{3} = {}^{?}\mathbf{?}$$

The multiplicand is $^{+}\mathbf{2}$ and the multiplier is $^{-}\mathbf{3}$.

The sign of the multiplicand states which side of the number line that the calculations will take place, which is the positive side. The multiplication operator states which way to face on the predetermined side of the number line, which is towards positive infinity. The multiplicand states how far to move on the number line per jump, and this is 2 positive units per jump. The multiplier states how many times the multiplicand moves repeatedly on the number line, which is 3 jumps. The sign of the multiplier states what environment the calculations are taking place, which is in a negative environment.

Hence: $^{+}\mathbf{2} \times {}^{-}\mathbf{3} = {}^{+}\mathbf{6}$

The negative environment defines the physical state of the multiplier or groups, without influencing the value of the multiplicand or its direction on the number line.

Use the following multiplication model for the third application:

$$^{-}\mathbf{2} \times {}^{+}\mathbf{3} = {}^{?}\mathbf{?}$$

The multiplicand is $^{-}\mathbf{2}$ and the multiplier is $^{+}\mathbf{3}$.

The sign of the multiplicand states which side of the number line that the calculations will take place, which is the negative side. The multiplication operator states which way to face on the predetermined side of the number line, which is towards negative infinity. The multiplicand states how far to move on the number line per jump, and this is 2 negative units per jump. The multiplier states how many times the multiplicand moves repeatedly on the number line, which is 3 jumps. The sign of the multiplier states what environment the calculations are taking place, which is in a positive environment.

Hence: $^{-}\mathbf{2} \times {}^{+}\mathbf{3} = {}^{-}\mathbf{6}$

Example 4: The Double Negative Reality

Use the following multiplication model for the final application:

$$^{-}2 \times {}^{-}3 \overset{?}{=} ?$$

The multiplicand is $^{-}2$ and the multiplier is $^{-}3$.

The sign of the multiplicand states which side of the number line that the calculations will take place, which is the negative side. The multiplication operator states which way to face on the predetermined side of the number line, which is towards negative infinity. The multiplicand states how far to move on the number line per jump, and this is 2 negative units per jump. The multiplier states how many times the multiplicand moves repeatedly on the number line, which is 3 jumps. The sign of the multiplier states what environment the calculations are taking place, which is in a negative environment.

Hence: $^{-}2 \times {}^{-}3 = {}^{-}6$

While a negative multiplied by a negative is traditionally accepted as a positive, this application demonstrates a different reality. By applying the same rigorous, consistent process used in every previous calculation, the true movement of this interaction is revealed. The negative environment of the multiplier represents a physical state of the groups or containers, but it does not possess the operational power to force the multiplicand across the zero barrier into positive territory. Because the multiplicand is the primary identity of the calculation, its sign dictates the territory. A negative object remains a negative object regardless of the environment it is placed in; therefore, repeatedly adding negative units in a negative space can only result in a greater amount of negative units.

Below are all four sign combination number sentences and their respective answers for how multiplication is done after applying the previous calculations.

$$^{+}2 \times {}^{+}3 = {}^{+}6$$

$$^{+}2 \times {}^{-}3 = {}^{+}6$$

$$^{-}2 \times {}^{+}3 = {}^{-}6$$

$$^{-}2 \times {}^{-}3 = {}^{-}6$$

The Inverse Bridge: Returning to Zero

If multiplication builds the territory through repeated addition, then division is the map that reveals the path back to zero through repeated subtraction. If this is true, then the inverse is also true: when division is performed first, multiplication takes the calculation back to the original starting point. This relationship ensures that the mathematical story remains consistent. Whether moving away from zero or returning to it, the identity of the numbers involved does not change. To maintain this consistency, the rules of the territory must be applied to division with the same exactness as they were to multiplication. How would multiplication of large numbers be done now?

$$\begin{array}{rl} 172 & \\ \underline{\times 8} & 2 \times 8 = 16 \\ 16 & 70 \times 8 = 560 \\ 560 & 100 \times 8 = 800 \\ \underline{+800} & \\ 1376 & \end{array} \qquad \begin{array}{rl} 172 & \\ \underline{\times {}^{-}8} & 2 \times {}^{-}8 = 16 \\ 16 & 70 \times {}^{-}8 = 560 \\ 560 & 100 \times {}^{-}8 = 800 \\ \underline{+800} & \\ 1376 & \end{array}$$

$$\begin{array}{rl} {}^{-}172 & \\ \underline{\times 8} & {}^{-}2 \times 8 = {}^{-}16 \\ {}^{-}16 & {}^{-}70 \times 8 = {}^{-}560 \\ {}^{-}560 & {}^{-}100 \times 8 = {}^{-}800 \\ \underline{+{}^{-}800} & \\ {}^{-}1376 & \end{array} \qquad \begin{array}{rl} {}^{-}172 & \\ \underline{\times {}^{-}8} & {}^{-}2 \times {}^{-}8 = {}^{-}16 \\ {}^{-}16 & {}^{-}70 \times {}^{-}8 = {}^{-}560 \\ {}^{-}560 & {}^{-}100 \times {}^{-}8 = {}^{-}800 \\ \underline{+{}^{-}800} & \\ {}^{-}1376 & \end{array}$$

This multiplication process can of course be done as normal, on one line, but with one added alteration that has not been seen before. When observing the negative examples ${}^{-}172 \times 8$ and ${}^{-}172 \times {}^{-}8$ the initial calculation of both is ${}^{-}16$; this requires placing a negative sign to the left, below the vinculum—the formal term for the line separating the calculation from the product. The 6 from the ${}^{-}16$ will sit on the far right under the vinculum and in line with the units. This is done in the same fashion as was done in Chapter 6 when a larger number was subtracted from a smaller number. For the calculation of $172 \times (8$ or ${}^{-}8)$, simply add the positive amounts. In the same manner, for ${}^{-}172 \times (8$ or ${}^{-}8)$, add the negative amounts to gain a result.

Now that the calculations have been corrected, it is possible to demonstrate how well division with negative numbers is performed. When approaching the division of larger values, a core principle must be maintained: the dividend occupies the same foundational role as the multiplicand. Just as the focus remains on the multiplicand during multiplication, it must be remembered that subtracting from a negative dividend will consistently yield a negative result. Within this framework, it is impossible to divide across the zero boundary towards a positive value. If wanting to attempt repeated subtraction, it will be the dividend subtracting the divisor amount, while using the sign of the dividend.

Putting the New Calculations to the Test

$$\begin{array}{r} {}^{+}98 \\ {}^{+}2\overline{\smash{)}\,{}^{+}196} \\ \underline{-{}^{+}18} \\ {}^{+}16 \\ \underline{-{}^{+}16} \\ 0 \end{array} \qquad \begin{array}{l} {}^{+}19 \div {}^{+}2 = {}^{+}9 \\ {}^{+}9 \times {}^{+}2 = {}^{+}18 \\ \text{Subtract } {}^{+}18 \\ {}^{+}16 \div {}^{+}2 = {}^{+}8 \\ {}^{+}8 \times {}^{+}2 = {}^{+}16 \\ \text{Subtract } {}^{+}16 \end{array}$$

$${}^{+}196 \div {}^{+}2 = {}^{+}98$$

$$\begin{array}{r} {}^{+}98 \\ {}^{-}2\overline{\smash{)}\,{}^{+}196} \\ \underline{-{}^{+}18} \\ {}^{+}16 \\ \underline{-{}^{+}16} \\ 0 \end{array} \qquad \begin{array}{l} {}^{+}19 \div {}^{-}2 = {}^{+}9 \\ {}^{+}9 \times {}^{-}2 = {}^{+}18 \\ \text{Subtract } {}^{+}18 \\ {}^{+}16 \div {}^{-}2 = {}^{+}8 \\ {}^{+}8 \times {}^{-}2 = {}^{+}16 \\ \text{Subtract } {}^{+}16 \end{array}$$

$${}^{+}196 \div {}^{-}2 = {}^{+}98$$

$$\begin{array}{r} {}^{-}98 \\ {}^{+}2\overline{\smash{)}\,{}^{-}196} \\ \underline{-{}^{-}18} \\ {}^{-}16 \\ \underline{-{}^{-}16} \\ 0 \end{array} \qquad \begin{array}{l} {}^{-}19 \div {}^{+}2 = {}^{-}9 \\ {}^{-}9 \times {}^{+}2 = {}^{-}18 \\ \text{Subtract } {}^{-}18 \\ {}^{-}16 \div {}^{+}2 = {}^{-}8 \\ {}^{-}8 \times {}^{+}2 = {}^{-}16 \\ \text{Subtract } {}^{-}16 \end{array}$$

$${}^{-}196 \div {}^{+}2 = {}^{-}98$$

$$\begin{array}{rl} ^{-}98 & \quad ^{-}19 \div {}^{-}2 = {}^{-}9 \\ ^{-}2\overline{)\,^{-}196} & \quad ^{-}9 \times {}^{-}2 = {}^{-}18 \\ \underline{-\,^{-}18} & \quad \text{Subtract } ^{-}18 \\ ^{-}16 & \quad ^{-}16 \div {}^{-}2 = {}^{-}8 \\ \underline{-\,^{-}16} & \quad ^{-}8 \times {}^{-}2 = {}^{-}16 \\ 0 & \quad \text{Subtract } ^{-}16 \end{array}$$

$$^{-}196 \div {}^{-}2 = {}^{-}98$$

At the start of this chapter, there were various ways in which using negatives in calculations did not work well, or not at all. Now those calculations can be checked to see if they work correctly with the new concepts. This verification demonstrates that the identity of the multiplicand remains the primary anchor of the result, regardless of the environment that the multiplier is in. By following these mechanical steps, the logical gaps of traditional arithmetic are exposed and corrected.

Multiplication With Positive and Negative Signs Working Together

A positive multiplied by a positive equals a positive result. This can only be calculated on the positive side of zero, moving towards positive infinity using repeated addition, regardless of the environment of the calculation.

$$^{+} \times {}^{+} = {}^{+}$$

A positive multiplied by a negative equals a positive result. This can only be calculated on the positive side of zero, moving towards positive infinity using repeated addition, regardless of the environment of the calculation.

$$^{+} \times {}^{-} = {}^{+}$$

A negative multiplied by a positive equals a negative result. This can only be calculated on the negative side of zero, moving towards negative infinity using repeated addition, regardless of the environment of the calculation.

$$^{-} \times ^{+} = ^{-}$$

A negative multiplied by a negative equals a negative result. This can only be calculated on the negative side of zero, moving towards negative infinity using repeated addition, regardless of the environment of the calculation.

$$^{-} \times ^{-} = ^{-}$$

Multiplication is the inverse of division.

Division with Positive and Negative Signs Working Together

A positive divided by a positive equals a positive result. This can only be calculated on the positive side of zero, moving towards zero using repeated subtraction, regardless of the environment of the calculation, and never becoming less than zero.

$$^{+} \div ^{+} = ^{+}$$

A positive divided by a negative equals a positive result. This can only be calculated on the positive side of zero, moving towards zero using repeated subtraction, regardless of the environment of the calculation, and never becoming less than zero.

$$^{+} \div ^{-} = ^{+}$$

A negative divided by a positive equals a negative result. This can only be calculated on the negative side of zero, moving towards zero using repeated subtraction, regardless of the environment of the calculation, and never becoming greater than zero.

$$^- \div {}^+ = {}^-$$

A negative divided by a negative equals a negative result. This can only be calculated on the negative side of zero, moving towards zero using repeated subtraction, regardless of the environment of the calculation, and never becoming greater than zero.

$$^- \div {}^- = {}^-$$

Division is the inverse of multiplication.

Investigation of Negative Numbers with Positive Exponents

Earlier, exponents increasing by 1 created an illogical pattern: positive, negative, positive, negative. The following illustrates the consistency of the new framework:

$$({}^-3)^2 = {}^-9$$
$$({}^-3)^3 = {}^-27$$
$$({}^-3)^4 = {}^-81$$
$$({}^-3)^5 = {}^-243$$

The pattern has been fixed and the calculations run smoothly.

The Group Identity Marker

To understand the repeated addition below, a distinction must be made between the Identity (the number inside the parentheses) and the Group Environment (the superscript outside). The superscript sign before each set of parentheses is a descriptor of the multiplier, identifying the nature of the group being added. A $^-$ outside the parentheses indicates a Negative Group, while a $^+$ indicates a Positive Group. These signs do not, at any time, stem from the existence of the operators, nor do they interact with them. They simply define the type of container being placed into the calculation, entirely separate from the mathematical actions being performed.

Exponents Broken Down into Repeated Addition:
Nothing needs to be forced to work as the logic follows a natural progression.

$$\begin{aligned} {}^{+}3^2 &= {}^{+}9 \\ &= {}^{+}3 \times {}^{+}3 \\ &= 0 + {}^{+}({}^{+}3) + {}^{+}({}^{+}3) + {}^{+}({}^{+}3) \end{aligned}$$

This is the same as before, which has not changed and works correctly. If exactly the same is done for negative numbers, it looks like the following example:

$$\begin{aligned} {}^{-}3^2 &= {}^{-}9 \\ &= {}^{-}3 \times {}^{-}3 \\ &= 0 + {}^{-}({}^{-}3) + {}^{-}({}^{-}3) + {}^{-}({}^{-}3) \end{aligned}$$

The complexity is removed, making the process straightforward and easy to understand. There is a significantly lower chance of misconception.

$$\begin{aligned} {}^{-}3^3 &= {}^{-}27 \\ &= {}^{-}3 \times {}^{-}3 \times {}^{-}3 \\ &= 0 + {}^{-}({}^{-}3) + {}^{-}({}^{-}3) + {}^{-}({}^{-}3) + {}^{-}({}^{-}3) + {}^{-}({}^{-}3) + {}^{-}({}^{-}3) \\ &\qquad + {}^{-}({}^{-}3) + {}^{-}({}^{-}3) + {}^{-}({}^{-}3) \end{aligned}$$

Further examples are not needed as all the forthcoming results will be negative, as they should be. The parentheses hold the multiplicand (the identity), while the superscript negative sign outside identifies the groups (the multiplier). This ensures that the calculations remain consistent across all operations.

Continuous Multiplication of Positive and Negative Values

When multiplying a constant by the previous result was tried earlier, it proved to be inconsistent, showing alternating negative and positive results. Now with new values:

$$3 \times 2 = 6$$
$$6 \times 2 = 12$$
$$12 \times 2 = 24$$
$$24 \times 2 = 48$$

As expected nothing has changed for basic positive multiplication.

$$^{-}3 \times 2 = {}^{-}6$$
$$^{-}6 \times 2 = {}^{-}12$$
$$^{-}12 \times 2 = {}^{-}24$$
$$^{-}24 \times 2 = {}^{-}48$$

These are exactly as they were at the start of the chapter, negatives have negative results.

$$3 \times {}^{-}2 = 6$$
$$6 \times {}^{-}2 = 12$$
$$12 \times {}^{-}2 = 24$$
$$24 \times {}^{-}2 = 48$$

Unlike earlier findings, this shows that a positive multiplicand multiplied by a negative amount will not change the final result.

$$^{-}3 \times {}^{-}2 = {}^{-}6$$
$$^{-}6 \times {}^{-}2 = {}^{-}12$$
$$^{-}12 \times {}^{-}2 = {}^{-}24$$
$$^{-}24 \times {}^{-}2 = {}^{-}48$$

This is also very different from the original results, and like above, the calculations must increase towards negative infinity as the numbers grow with every multiplication.

Negative Logarithms

Now that positive and negative values work together correctly, calculating logarithms is much easier and has a much greater understanding:

$$\log_{-3} {}^{-}27 = 3$$

This works consistently because the base and the argument are the same identity; therefore, both must have the same sign. For example: the base, $^{-}3$, is an amount of something, and the argument, $^{-}27$, is the total amount of that same something after the calculations are finished. The exponent, 3, on the other hand is an operator or more precisely is a multiplier.

Because the product of two negatives in traditional mathematics results in a positive, negative logarithms cannot work in the old system. Under those rules, both equations below would have no real solutions:

$$\log_{-2} {}^{-}4 = ?$$

$$({}^{-}2)^{?} = 4$$

However, in this framework, if the log of an argument is negative, the base must be negative. If the log of an argument is positive, then the base must be positive:

$$\log_{-2} {}^{-}4 = 2$$

$$\log_{2} 4 = 2$$

The exponent, whether the base and arguments are positive or negative, will always be positive. This is because it is an operator and it cannot change the value the original entity.

$$(^{-}2)^2 = {}^{-}4$$

$$(2)^2 = 4$$

Beyond the Imaginary: A Unified Field of Arithmetic

This concludes a very long and vital chapter. These findings provide a definitive answer for anyone who has ever questioned the logic of two negative ideas forming a positive conclusion. By aligning basic arithmetic with the Natural Order, the glitches of the past are left behind in favour of a mathematically consistent future. While traditional axioms have long ignored the negative identity, this work establishes a unified framework where the negative side of the number line is finally given its own territorial integrity.

In this model, the square root of a negative number is no longer an impossible or imaginary concept. It is simply the inverse of squaring a negative value. Since a negative multiplicand multiplied by a negative multiplier remains in negative territory, the square root of a negative number is naturally a negative value. The calculation never leaves its original side of the number line, maintaining its absolute identity. To answer the question posed at the start of this chapter: Is the square root of a negative number a physical possibility? The answer is an absolute yes; the square root of a negative number is a simple reality.

$$\sqrt{^{-}9} = {}^{-}3$$

Everything presented in this chapter is based on the foundational assumption that a positive number added to a positive number equals a positive. With this Natural Order now running smoothly—free of errors or concepts that fail—it is possible to perform calculations on the negative side of the number line with the same precision as the positive side.

8

Negative Division with Remainders: The Last Hurdle

Through rigorous testing of the framework of the previous chapter, a process was discovered that fails to align with consistent mathematical logic: the remainders of negative division. Though seemingly insignificant, this concept represents a major structural flaw. The reason this issue appears after Chapter 7 is that it was not fully identified until all prior calculations in that chapter were completed.

To find a solution, the current perception of these calculations must be examined. While dividing negative quantities evenly is simple, an issue arises when remainders appear, as the old system often allows for two different results. Strangely, standard practice suggests it is acceptable to choose whichever result suits a specific need—a subject of debate that lacks a definitive, mechanical answer. Usually, it is accepted that a decimal remainder will be negative, yet Euclidean division insists the remainder must be positive. An example of these conflicting remainders is seen below (where the first is standard and the second is Euclidean):

$$^{-}8 \div 3 = {}^{-}2 \text{ with } r^{-}2$$

$$^{-}8 \div 3 = {}^{-}3 \text{ with } r1$$

In reality, there cannot be two different answers for such a simple concept; and that truth is absolute. While mental arithmetic for small negative quotients is simple, the remainder is more confronting. Euclidean division is a method devised to find these remainders, yet it is a patch that utilises multiplication as the main operator rather than true division. Another common workaround is simply performing positive division and replacing signs where needed—an imprecise approach that relies on guessing rather than mathematical certainty.

It is time to identify what causes this mystery and see if a solution exists to fix the glitch. In standard division, it is agreed that $^{-}17$ of something divided by 2 is $^{-}8$ with $^{-}1$ left over, the remainder. This of course will give 2 lots of $^{-}8$ with a bit extra. This can be verified by calculating the inverse of division, which is multiplication:

$$^{-}8 \times 2 + {}^{-}1 = {}^{-}17$$

However, with Euclidean division that logic changes as the quotient jumps from $^{-}8$ to $^{-}9$ even though the dividend is not $^{-}18$:

$$^{-}17 \div 2 = {}^{-}9\, r\, ^{?}?$$

The core of the issue is that standard negative division divides towards zero. This leaves a remainder $^{-}1$, which is technically greater in value (closer to zero) than the starting dividend $^{-}17$. Therefore it is considered that it cannot possibly be the correct answer because there cannot be a greater value as a remainder than the original starting amount. Although when logic steps in, it would appear that $^{-}1$ as a remainder is the correct answer because this is what is left over after performing standard division. According to Euclidean division, the remainder would be 1 such that:

$$^{-}9 \times 2 + 1 = {}^{-}17$$

This of course makes sense and seems perfectly logical.
After going back over all the information, including the first to last drafts of this book, there was no obvious reason why this negative division with a remainder did not provide a unique result. Upon closer inspection, it is clear to see that there are two concepts happening here at once. Numbers are counting up as they move left on the number line and each number is considered to be less than the previous number as they move in the same direction, which is towards negative infinity on the number line. There is a very peculiar inconsistency happening here and the number line seems to be the next place to investigate. By applying comparative reasoning, we can determine exactly where the system breaks down.

The Number Line and What is Well Known About it

The positive side of the number line and how it is used and visually perceived...
The following outlines how the positive side of the number line is used and what is known to be fact:

1. Negative numbers do not exist on the positive side of the number line.
2. Starting from zero or any positive number, counting individual positive quantities moves towards positive infinity; this movement is to the right.

3. Starting from zero or any positive number, positive quantities that are added together move towards positive infinity; this movement is to the right.
4. Positive non-zero multiplicands that are multiplied by positive non-zero multipliers move towards positive infinity; this movement is to the right.
5. Starting from any positive number, counting back individual positive quantities moves towards zero; this movement is to the left.
6. Starting from any positive number, positive quantities subtracted from that number move towards zero; this movement is to the left.
7. Positive non-zero dividends that are divided by positive non-zero divisors move towards zero. This movement is to the left and can never pass through zero to the negative side of the number line.
8. The concept of numbers and the concept of positives are two separate ideas.
9. Actual physical numbers increase in value as they move from zero towards positive infinity.
10. The concept of positive increases in value as it moves from zero towards positive infinity.
11. Actual physical numbers decrease in value as they move towards zero on the positive side of the number line.
12. The concept of positive decreases in value as it moves towards zero on the positive side of the number line.

The negative side of the number line, which must be the complete opposite of the positive side of the number line and includes the new concepts from this book...

1. Positive numbers do not exist on the negative side of the number line.
2. Starting from any negative number including zero, counting on individual negative quantities moves towards negative infinity, this movement on the number line is to the left.
3. Starting from any negative number including zero, negative quantities that are added together move towards negative infinity, this movement on the number line is to the left.
4. Negative non-zero multiplicands that are multiplied by negative non-zero multipliers move towards negative infinity (new concept), this movement on the number line is to the left.
5. Starting from any negative number, counting back individual negative quantities moves towards zero, this movement on the number line is to the right.
6. Starting from any negative number, negative quantities that are subtracted from that negative number move towards zero; this movement on the number line is to the right.
7. Negative non-zero dividends that are divided by negative non-zero divisors (new concept), move towards zero. This movement on the number line is to the right and can never pass through zero to the positive side of the number line.
8. The concept of numbers and the concept of negative are two separate ideas.
9. Actual physical numbers increase in value as they move towards negative infinity from zero on the negative side of the number line.
10. The concept of negative decreases in value as it moves towards negative infinity from zero on the negative side of the number line.

11. Actual physical numbers decrease in value as they move towards zero on the negative side of the number line.
12. The concept of negative increases in value as it moves towards zero on the negative side of the number line.

The Distinction Between Opposite and Inverse

The information presented for the positive and negative sides of the number line is vital because it demonstrates how these two concepts should be exact opposites, not inverses.

Points 10 and 12 from the negative side—"the concept of negative decreases..." and "the concept of negative increases..."—are very interesting because they do not act in an opposite way compared to the same points on the positive side, which they should. What is even more interesting is how the points 9 and 10 on the positive side move in unison towards positive infinity. However on the negative side, point 9 (physical numbers) increases towards negative infinity while point 10 (the concept) is the only one of those four points to decrease while heading towards negative infinity. This structural divergence is the root of the glitch; it is where standard mathematical concepts begin to fail, necessitating the various patches and axiomatic assumptions currently used to force a result.

The Number Line Perspective

Considering if the number line had an issue was even more unexpected than it was when finding a problem with negative division and its remainders. After looking at the number line and all the information that is currently known, it seemed that changing the number line was not possible as the numbers worked perfectly, but a search was made nonetheless.

The way in which a solution was found was so incredibly simple. It required thinking of negative not as less than the previous number but more; for instance, increasing negative numbers can make something more amounts of negative. The easy part about negatives is that not many things can be construed as negative. For instance, negative apples or objects cannot really be used, but negatives can be used in a more common way such as debt. Temperature and movement can also be applied. There of course may be many others, but these are adequate for the needs of this investigation.

The following is quite difficult to see at first because of what is already known and understood to be true, so it may take some time to see the concept. It does, however, make all negative concepts work properly and in particular negative division with remainders. Describing how a negative number is less than another negative number using a number line is easy, but to describe actual real-life objects is not so simple. The reason for this is because the word "more" is used in the wrong context.

- Traditional View: Moving from $^{-}2$ to $^{-}7$ is seen as five less than negative two.
- Identity View: Moving from $^{-}2$ to $^{-}7$ is more negative. There is a greater amount of negative territory, and can be seen as five more than negative two.

Consider a car loan of $\$10{,}000$ ($^{-}10{,}000$). If a second loan for a boat is taken for $\$6{,}000$, the result is:

$$^{-}10{,}000 + {}^{-}6{,}000 = {}^{-}16{,}000$$

In everyday language, this is more debt. On the traditional number line view, $^{-}16{,}000$ is less than $^{-}10{,}000$, which suggests that there is less debt. This clearly does not make sense.

By acknowledging that negatives become greater in magnitude, the numbers and the concept of negative can now increase towards infinity together. This shift restores the fundamental logic of negative division and subtraction. As values are subtracted towards zero both the physical numbers and the concept of negatives also decrease in unison. Within the previous system, taking an amount away from a large negative value resulted in a larger number— a logical impossibility. By defining the negative direction as an increase in negative amount, subtraction and division, particularly division with remainders, finally operate with mechanical precision.

For negative division there can now only be one answer for the remainder:

$$^{-}17 \div 6 = {}^{-}2\,r\,{}^{-}5$$

The last hurdle has been cleared. By aligning the language of mathematics with the physical reality of the world, the Natural Order of numbers has been restored.

Now that the negative identity is stabilised and the number line reflects the reality of increasing magnitude, previously impossible concepts can be approached with logical consistency. The restoration of this order allows for the exploration of:

- **Negative Primes:** Identifying numbers within the negative identity that have no factors other than themselves and $^{-}1$.
- **Negative Perfect Numbers:** Observing the symmetry of numbers whose divisors sum to the number itself.

- **Square Roots of Negative Numbers:** Calculating the root of a negative square as a self-contained environment, removing the need for imaginary placeholders.

9

A Product of Two Negatives: Misconceptions and Reality

Proofs and Justifications Explained

Ever since it was first proposed that a negative multiplied by a negative equates to a positive outcome, there has been a continuous effort to find inventive ways to prove the concept. This has led to some well-thought-out proofs and some quite quirky ones as well. None of these have been easy to formulate, and the question must be asked: why has this concept been so difficult to prove? The simple answer is, because the goal has been to prove something that was incorrect in the first place.

Most proofs rely on a simple statement to be true: the concept of adding a negative, $+(-)$. This generally occurs in number sentences that look like this:

$$\boldsymbol{a} + (-\boldsymbol{a}) = \mathbf{0}$$

As established earlier in this book, the additive inverse—or adding a negative—cannot be achieved on the positive side of the number line. The hyphen or dash $(-\boldsymbol{a})$ is often used interchangeably as either a subtraction operator or a negative sign whenever a change is needed to make a proof function.

Another way of accepting a proof is the inclusion of axioms, corollaries, and other known number facts. These ideas do not need to be detailed here; they are simply accepted as facts. However, a significant problem exists: there do not seem to be any official axioms for negative concepts. Above all else, it remains incredibly difficult to provide a real proof that the product of two negatives equals a positive. The following proofs are just a few of many, but most share the same or similar general ideas.

Proof 1

The following is a standard version of a proof found in textbooks worldwide:

$$3 \times 0 = 0$$
$$-3(4 + (-4)) = 0$$
$$-3 \times 4 + -3 \times -4 = 0$$
$$-12 + 12 = 0$$

Therefore, it is claimed that -3×-4 must equal $+12$, concluding that a negative multiplied by a negative equals a positive.

While this looks perfect on the surface, it relies entirely on a specific notation. In early mathematics, adding a negative is taught as being synonymous with subtraction:

$4 + (-4)$ is the same as $(4 - 4)$

However, after inserting positive and negative superscripts, if the proof is attempted using standard subtraction, the negative × negative disappears:

$${}^{-}3 \times 0 = 0$$
$${}^{-}3({}^{+}4 - {}^{+}4) = 0$$
$${}^{-}3 \times {}^{+}4 - ({}^{-}3 \times {}^{+}4) = 0$$
$${}^{-}12 - {}^{-}12 = 0$$

This still results in zero, but it fails to produce a negative multiplied by a negative. The textbook proof requires the adding a negative concept specifically to force the two negatives together. The fundamental error in the first proof is that a negative value cannot be added to a space where no negative values exist. As established by the Natural Order, adding a negative can only occur on the negative side of the number line where those values already exist. Attempting this proof using the Identity Framework, where the multiplicand is written first and the multiplier sits outside the parentheses, a different pattern emerges:

$$0 \times {}^{-}3 = 0$$
$$^{-}3\left(0 + {}^{-}4 - ({}^{-}4)\right) = 0$$
$$(0 \times {}^{-}3) + ({}^{-}4 \times {}^{-}3) - ({}^{-}4 \times {}^{-}3) = 0$$
$$0 + {}^{-}12 - ({}^{-}12) = 0$$

In this calculation:

$$^{-}3 \times {}^{-}4 = {}^{-}12$$

Under this logic, a negative multiplied by a negative equals a negative. Even though this second calculation runs smoothly, it must be acknowledged that this is not a proof in the absolute sense—it is patterning. The second line inside the parentheses is essentially the same as the fourth line; it is simply the result of being multiplied by $^{-}3$ and rewritten. In both the traditional textbook version and this identity-based version, the result is dictated by the pattern established in the setup. The proof is merely a reflection of the initial assumptions.

Proof 2: The Variable Substitution Trap

Another common proof found in mathematical forums uses variable substitution to force a positive result. The logic typically follows this path:

1. $({}^{-}a)({}^{-}b) + ({}^{-}ab) = ({}^{-}a)({}^{-}b) + ({}^{-}a)b$
2. $({}^{-}a)({}^{-}b) + ({}^{-}ab) = ({}^{-}a)({}^{-}b + b)$
3. $({}^{-}a)({}^{-}b) + ({}^{-}ab) = ({}^{-}a)(0)$
4. $({}^{-}a)({}^{-}b) + ({}^{-}ab) = 0$
5. $({}^{-}a)({}^{-}b) = ab$

This proof only functions if there is no recognised distinction between the multiplicand and the multiplier. In this string, the multiplier is the common variable (${}^{-}a$), while (b) is treated as both a negative and a positive, which assigns it two different functional values.

Traditional Numerical Example:

Using the traditional Negative × Negative = Positive assumption:

a) $({}^{-}2)({}^{-}3) + ({}^{-}2 \times {}^{+}3) = ({}^{-}2)({}^{-}3) + ({}^{-}2){}^{+}3$
b) $({}^{-}2)({}^{-}3) + ({}^{-}6) = ({}^{-}2)({}^{-}3 + {}^{+}3)$
c) $({}^{-}2)({}^{-}3) + ({}^{-}6) = ({}^{-}2)(0)$
d) $({}^{-}2)({}^{-}3) + ({}^{-}6) = 0$
e) $({}^{-}2)({}^{-}3) + {}^{-}6 - ({}^{-}6) = -({}^{-}6)$
f) $({}^{-}2)({}^{-}3) = {}^{+}6$

The Identity Framework Correction:

However, if the multiplicand remains true to its identity as a negative quantity, the calculation proves that the product of two negatives remains negative:

a) $(^{-}2)(^{-}3) + (^{-}2 \times {}^{+}3) = (^{-}2)(^{-}3) + (^{-}2)^{+}3$

In line b), the multiplier $(^{-}2)$ is factored out of the equation on the RHS, clearly revealing it as the consistent force acting upon the parentheses. $(^{+}3)$ was the initial multiplier on the LHS but as stated $(^{-}2)$ is the true multiplier on both sides.

b) $(^{-}2)(^{-}3) + (^{+}6) = (^{-}2)(^{-}3 + {}^{+}3)$
c) $(^{-}2)(^{-}3) + (^{+}6) = (^{-}2)(0)$
d) $(^{-}2)(^{-}3) + (^{+}6) = 0$
e) $(^{-}2)(^{-}3) + {}^{+}6 - (^{+}6) = -\ (^{+}6)$
f) $(^{-}2)(^{-}3) = {}^{-}6$

By maintaining the negative identity of the result in step (b), the final result (f) correctly identifies the product as negative. The traditional proof only works by switching the identity of the product in the middle of the calculation to satisfy the goal of reaching zero. If a simplified version is needed, the second part of this proof is compelling. The variable substitution trap is a logical loop. It assumes that $(^{-}a)b$ is the correct way to describe the environment and uses that assumption to prove itself.

$$(^{-}a)(^{-}b) + (^{-}ab) = (^{-}a)(^{-}b) + (^{-}a)b$$

$$(^{-}a)(^{-}b) + (^{-}ab) = (^{-}a)(^{-}b) + (^{-}ab)$$

Since both sides of the initial equation are identical, the proof is revealed as a choice rather than a necessity. If $(^{-}a)(^{-}b)$ on the LHS is assigned as positive, the RHS must follow. If it is assigned as negative, the RHS remains negative. The logic of this framework maintains the path that respects the Identity of the number, ensuring the calculation remains territorially consistent from start to finish.

Proof 3: The Properties of Distribution

The distributive property is often cited as the ultimate proof that two negatives must equal a positive. However, this is only true if the identity of the multiplicand is ignored. Following the traditional rule (Negative × Negative = Positive), the distribution appears to work:

$$^{-}1(^{+}1 + {}^{-}1) = (^{-}1 \times {}^{+}1) + (^{-}1 \times {}^{-}1)$$
$$^{-}1(0) = {}^{-}1 + {}^{+}1$$
$$0 = 0$$

It is often argued that if the product of two negatives were negative, distribution would fail. Changing the result to negative without adjusting the framework leads to an incorrect result:

$$^{-}1(^{+}1 + {}^{-}1) = (^{-}1 \times {}^{+}1) + (^{-}1 \times {}^{-}1)$$
$$^{-}1(0) = {}^{-}1 + {}^{-}1$$
$$0 = {}^{-}2$$

This error occurs because the multiplicand has not been acknowledged. If the negative × negative rule is changed without maintaining the consistency of the other results (such as positive × negative = positive), the balance is lost.

Distribution works exactly as intended when the multiplicand remains the focus. Although earlier chapters established the multiplicand as the first mention throughout this book, placing the multiplicand inside the parentheses is acceptable provided its location is known at all times. In the example below, the multiplicand is highlighted to reveal the consistent identity:

$$^{-}1(^{+}\mathbf{1} + {}^{-}\mathbf{1}) = (^{-}1 \times {}^{+}\mathbf{1}) + (^{-}\mathbf{1} \times {}^{-}1)$$
$$^{-}1(0) = {}^{+}1 + {}^{-}1$$
$$0 = 0$$

By maintaining the identity of the multiplicand, the equation balances perfectly. This confirms that a product of two negatives equals a negative while remaining mathematically sound.

The use of adding a negative in this specimen is included strictly to align with the traditional phrasing of the distributive law. While this book typically dismisses such academic noise in favour of a more direct physical transit, it is presented here as the world expects to see it to prove that even within the old framework; the identity of the multiplicand remains the absolute truth.

The Multiplication Area Model Proof

One example of proving that multiplying two negatives to produce a positive is taking a square and dissecting into nine smaller squares. There are numbers in the outer boxes that multiply in a criss-cross fashion to fill the four inner boxes, which is an area model:

$\times$	9	-3
7	63	-21
-4	-36	?

This proof works by multiplying through each square across or down the area model and filling the empty spaces with the correct answers or, more precisely for this proof, the correct sign:

$$7 \times 9 = 63$$

$$-4 \times 9 = -36$$

$$7 \times -3 = -21$$

$$-4 \times -3 = ?$$

For the area model above, it seems that the only possible correct answer is a positive because there are already two negatives and one positive. The standard model imposes an arbitrary rule: to maintain balance, the grid must contain an even number of positive and negative signs. Under this assumption, the missing square *must* be positive to satisfy the pattern.

$$-3 \times -4 = 12$$

After working out the correct answer with the correct sign, the four products are calculated together:

$$63 - 36 - 21 + 12 = 18$$

Then, the calculations on the outside are done to verify the correctness of calculations:

$$9 - 3 = 6 \text{ and } 7 - 4 = 3$$

These two results are then multiplied together:

$$6 \times 3 = 18$$

The result here matches perfectly to the calculations of all the products within the square. This concept seems to instantly confirm that two negatives must make a positive, as the empty square must be a positive. To find out the true nature of this simple but amazing square, discovering what an area model is and what it can be used for looks to be the most logical place to

start. Finding out what makes the area model function the way it does should help to reveal whether it is a real proof or not.
Learners are introduced to basic area models to help them discover what multiplication is all about. The process above shows how the area model accesses addition and subtraction to help work out answers of multiplication problems with large numbers. It also shows how positive and negative values, when united within the area model, are kept balanced by only allowing two of each sign, but at times may have all signs being the same. Positive and negative values must always be even in order to obtain a correct answer. Using all the values in the correct manner can help learners to solve problems that have large numbers, such as finding the area of a rectangle and in other areas of mathematics. This concept can be used at higher levels of mathematics.

Looking at the calculations from above, it is clear that there is a distinct pattern appearing. The numbers on the side of the box are both multiplied by the first number on top. Then, those same numbers on the same side are both multiplied again, but by the second number on top. These number combinations and the use of multiplication, addition, and subtraction together are consistent with another related concept in mathematics called distributive problems. Another concept that has been used within the above calculations is that the subtraction symbols are being used as negative signs. From the information gathered so far in this book and showing how the subtraction symbols should be used, it appears that the negative signs—or lack thereof—may have played a role in how the calculations were interpreted and ultimately used for the proof. Using the ideas of the distributive properties to look at the calculations from the area model, it appears that those calculations can be displayed in a more orderly fashion. Assuming the reader already knows what the distributive properties are and how they are used; it is time to examine those numbers more closely.

The first task to take care of is to identify the multiplicands and the multipliers. This can be done in any order—for example, starting with the 7 and -4 or the 9 and -3. The seven and the negative four will be selected as the multipliers, which leaves the nine and the negative three as the multiplicands. These will be placed into the form of distributive problems with the multipliers outside of the parentheses. To give greater clarity and understanding, all the calculations will be done with correct placement of the negative signs for the appropriate numbers:

$$\begin{aligned} 7(9 + {}^{-}3) &= (9 \times 7) + ({}^{-}3 \times 7) \\ &= 63 + {}^{-}21 \\ &= 42 \end{aligned}$$

$$\begin{aligned} {}^{-}4(9 + {}^{-}3) &= (9 \times {}^{-}4) + ({}^{-}3 \times {}^{-}4) \\ &= {}^{-}36 + 12 \\ &= {}^{-}24 \\ 42 + {}^{-}24 &= 18 \end{aligned}$$

Even though the area model shows that the nine and negative three together are being used as subtraction, a negative is not subtraction, so addition of the negative number must be used. After calculating the results of each distribution problem, the two are summed together to get a final result. These two distributive problems reveal three major logical violations that undermine the validity of the proof:

1. **The Adding a Negative Contradiction**

2. **The Misapplication of the Area Model**
3. **The Operator Deception**

Audit 1: The Adding a Negative Contradiction

The adding a negative issue has been discussed quite a bit earlier in this chapter and other chapters and still creates the same problem of not working properly, especially for a negative multiplied by a negative proof.

Audit 2: The Misapplication of the Area Model

When using area models at beginner levels of mathematics, multiplication is done by decomposing large numbers into smaller, more manageable numbers. This means that each decomposed number uses addition on the outer sides of the area model. For example, thirty-seven multiplied by eighteen: thirty-seven would be broken down into two numbers, thirty and seven, as would the eighteen using ten and eight. Calculating these numbers would not normally see them decomposed using subtraction. To achieve subtraction, thirty-seven multiplied by eighteen could be done as forty take away three and twenty take away two. This is how the above area model of six multiplied by three has been done, and area models are not really done in this way, they can be, but for new learners that would be detrimental to their learning.

Algebra, on the other hand, uses area models quite well with subtraction and negative values—for example, when looking into the basics of quadratic equations. In the form of factorising, $(x-3)(x-2)$, being displayed in an area model, the necessary information can be seen more clearly and gives a good visual representation of the quadratic equation. The area model above can be factorised in the same way as factorising a quadratic equation but fails to provide foundational proof. It merely demonstrates how algebra uses the area model as a visual container for pre-existing rules, rather than as a source of logical truth.

Audit 3: The Operator Deception

The biggest issue with this proof is how the subtraction operators are being used as negative signs. Looking at how the square came to be in the first place sees the multiplication problem that was posed, which is six multiplied by three. The numbers on the outer sides of the square use subtraction and are not negative numbers. When the calculations are performed from inside the area model, it is difficult to see what is really happening, but when displayed as distributive problems, visualising subtraction is much easier to understand. It also shows how it should be done, with subtraction:

$$7(9-3)-4(9-3)$$

The two distributive expressions above have the second being subtracted from the first, which is how the original decomposition of six multiplied by three was done (nine take away three and seven take away four):

$$\begin{aligned} 7(9-3) &= (9\times 7)-(3\times 7) \\ &= 63-21 \\ &= 42 \end{aligned}$$

Subtract the following from the first:

$$\begin{aligned} 4(9-3) &= (9\times 4)-(3\times 4) \\ &= 36-12 \\ &= 24 \end{aligned}$$

The final result:

$$42-24=18$$

This is not how distributive problems are done when performing everyday mathematics, but this is how it looks without shortcuts. The above is written to emphasise the subtraction of one distribution problem from another simply because both multipliers are positive. The shortcuts mainly refer to using subtraction as negative values in order to speed up the process, which does work very nicely. In conclusion, this clearly shows that the area model proof is not a proof in any way. Unless there are actual negative numbers in the distributive number sentences, the answer will not be negative, and this does include algebra distribution as well.

The outer numbers are all even and now show the operation of subtraction. The values within each square are all positives as well; the operations shown in each square are:

$\times$	9	-3
7	$+63$	-21
-4	$+36$	-12

Now, if these were construed as positive and negative values, it is consistent with the new combinations of positive and negative value i.e. negative multiplied by a negative equals negative.

Abstract and Subtraction Proofs

There are many obscure and abstract methods used to prove that two negatives equal a positive. These range from physical analogies—such as walking forward, then backward, then backward again while facing forward—to the use of positive and negative discs. In many cases, these stories rely on the concept

of cancelling out, which is a logical convenience rather than true subtraction.
Troubling analogies intended to mirror mathematical logic are also frequently encountered, such as, "If I say, 'Do not not eat'..." or "The enemy of your enemy is your friend..."
While these may be interpreted as correct in a certain linguistic or social context, they only result in a positive because the author intends for them to be interpreted that way. They are not absolute truths. These are social metaphors, not mathematical proofs.

A significant number of proofs rely on having two negatives in the form of a negative sign and a subtraction symbol e.g. $-(^{-}a)$. It must be stated clearly: these are not proofs of multiplication. The subtraction of a negative does move in a positive direction on the number line, but this is an operational movement, not the product of two negatives. Multiplying two identities is a different functional process than subtracting a value from a position. If a proof requires switching between a sign and an operator just to reach a positive result, it is not a proof, it is a manipulation of notation.

The Persistence of Identity in Division

A fundamental truth of the Natural Order is that if a positive number divided by itself remains positive, then a negative number divided by itself must remain negative. The process of division cannot change the inherent identity of the original value.

The Freezer Analogy (Positive Dividends)

Consider two positive entities, two apples $(^{+}2)$ being divided into two groups. Whether the environment is negative (a freezer) or positive (an oven), the entities remain apples. The focus is on the substance, not the temperature. Unless the mathematical story specifically shifts the focus to the temperature—making it the multiplicand—the identity of the entity as an apple remains unchanged by the divisor.

The Bank Account Analogy (Negative Dividends)

Similarly, a debt of two dollars ($^{-}2$) divided into two accounts remains a debt. One account is already a debt (negative account), one dollar debt divided into this account means that there is more debt and the identity of this dividend is still negative. Second account: if one dollar of that debt is moved into an account with a positive balance (positive account), the action of introducing that debt creates a lesser amount within that environment. The debt cannot magically transform into a positive asset simply by being moved; the identity of the dividend as debt is preserved across all groups.

The Misconception of a Number Divided by Itself

The phrase dividing a number by itself is often misleading. For example, $1 \div 1$ does not mean one apple divided by one apple. It means one something (the dividend) is being divided into a group of the same amount (the divisor). It is vital to remember that division is a process of distribution, not a fraction.

With an understanding of how division functions within the Natural Order, it becomes clear that division—even with negative numbers—only subtracts towards zero. It never passes through zero to the opposite side. The only way to move from positive to negative (or vice versa) is if the mathematical story provides specific extra information that allows for such a transition.

The Failure of Commutation in Negatives

Since the multiplicand and the dividend are the sole focus of any mathematical story, commutation does not work for either concept when one of the values is negative. A negative amount of something is fundamentally different from a positive amount; they cannot be treated as interchangeable. Performing commutation for like signs can still be done in theory as a shortcut, provided the nature of the multiplicand remains the focus at all times—especially in contexts where its identity is vital to the result. The dividend is the anchor of the problem. Therefore, a negative dividend divided by a positive divisor is never the same as a positive dividend divided by a negative divisor. The results are different because the starting identities are different; this is also true for multiplication:

$$3 \div {}^{-}3 \neq {}^{-}3 \div 3 \qquad 3 \times {}^{-}3 \neq {}^{-}3 \times 3$$

$$3 \div {}^{-}3 = 3 \qquad 3 \times {}^{-}3 = 3$$

$$^{-}3 \div 3 = {}^{-}3 \qquad {}^{-}3 \times 3 = {}^{-}3$$

The Successor Function for Negative Integers

A formal definition of addition is constructed through theorems, postulates, and deductive reasoning. One such foundational postulate is the successor function, complemented by its inverse, the predecessor function. Notably, Peano's axioms—the bedrock of arithmetic—are restricted to natural numbers and exclude negative values. The question of *why* they are excluded finds its resolution in the preceding chapter; however, a general overview of this concept for the operation is necessary first.

In its simplest form, the addition definition utilises the successor function to identify the next number. For example, the successor of five is:

$$5 + 1 = 6$$

Conversely, the predecessor function identifies one less by subtracting one natural number at a time, acting as the direct inverse of the addition definition:

$$7 - 1 = 6$$

These concepts are critical to understanding why robust axioms for negative numbers have not been established. Currently, the successor function for negatives calls for the next higher natural number and higher negative numbers are moving to the right on the number line. For negative five, this is written as:

$$^{-}5 + {}^{+}1 = {}^{-}4$$

For the predecessor of negative eight, the next lower natural negative number is to the left on the number line such that:

$$^{-}8 - {}^{+}1 = {}^{-}9$$

These calculations do not work logically within a unified axiom system because a positive number must be called to achieve the desired result. This is why negatives have historically been excluded. To fix the axiom for negatives, the successor must move left (towards negative infinity) and the predecessor must move right (towards zero). This movement is the direct opposite of the positive side.

The Corrected Successor Function

Under the corrected concept of the number line, where negative values increase in magnitude towards negative infinity, the functions now work properly:

The successor of negative five with the correct information added:

$$^{-}5 + {}^{-}1 = {}^{-}6$$

The predecessor of negative eight with the correct information added:

$$^{-}8 - {}^{-}1 = {}^{-}7$$

Verifying the S Function

Peano's axioms conclude that $\boldsymbol{S(0) = 1}$. Since this is traditionally done using positive values, it has not been clear which parts of the definition are positive. If the successor of 0 is 1, $\boldsymbol{S(0)} = {}^{+}\mathbf{1}$, then the function is calling a positive unit. The positive identity can be clarified as follows:

- $\boldsymbol{{}^{+}2 + {}^{+}1 = {}^{+}2 + S(0) = S({}^{+}2 + 0) = S({}^{+}2) = {}^{+}3}$
- $\boldsymbol{{}^{+}2 + {}^{+}2 = {}^{+}2 + S({}^{+}1) = S({}^{+}2 + {}^{+}1) = S({}^{+}3) = {}^{+}4}$
- $\boldsymbol{{}^{+}2 + {}^{+}3 = {}^{+}2 + S({}^{+}2) = S({}^{+}2 + {}^{+}2) = S({}^{+}4) = {}^{+}5}$

The Negative Successor Implementation

As the number line has been rectified to see the negative concept increasing towards negative infinity, the successor function can now be utilised as a negative function:

- $\boldsymbol{{}^{-}2 + {}^{-}1 = {}^{-}2 + S(0) = S({}^{-}2 + 0) = S({}^{-}2) = {}^{-}3}$
- $\boldsymbol{{}^{-}2 + {}^{-}2 = {}^{-}2 + S({}^{-}1) = S({}^{-}2 + {}^{-}1) = S({}^{-}3) = {}^{-}4}$
- $\boldsymbol{{}^{-}2 + {}^{-}3 = {}^{-}2 + S({}^{-}2) = S({}^{-}2 + {}^{-}2) = S({}^{-}4) = {}^{-}5}$

The Multiplication Successor Implementation

The beauty of the rectified successor function for negatives lies in its elegant simplicity; the multiplication definition for negative values now follows without error:

- $\mathbf{{}^{-}3 \cdot {}^{-}2 = {}^{-}3 \cdot \mathit{S}({}^{-}1) = {}^{-}3 + ({}^{-}3 \cdot {}^{-}1) = {}^{-}3 + {}^{-}3 = {}^{-}6}$

A formal proof for negative addition and negative multiplication has never been able to be done simply because the perceived use of the number line—specifically the negative side—was incorrect. That can now change. By using the Natural Order, these proofs are no longer quirky or inventive workarounds; they are the exact, logical opposites of positive proofs.

The Final Axiom: Symmetrical Successors

The formulation of these axioms was a long and deliberate process, ensuring they were perfect in every way. In this model, the successor function (S) is no longer a one-way street. It is a symmetrical expansion:

- On the **Positive side,** $S({}^{+}n)$ calls the next positive unit, moving further from zero.
- On the **Negative side,** $S({}^{-}n)$ calls the next negative unit, moving further from zero.

This maintains the Identity of the objects being counted. Whether counting positive apples or negative apples (debt), the act of succeeding always results in a greater magnitude of that specific identity.

An Amazing Pattern has Naturally Evolved

The following is a simple pattern which can be done without relying on any other justifications, more importantly; it makes the impossible, possible.

It is of course as natural as counting a whole bunch of apples...

$$5 \times 5 = 25$$
$$4 \times 4 = 16$$
$$3 \times 3 = 9$$
$$2 \times 2 = 4$$
$$1 \times 1 = 1$$
$$0 \times 0 = 0$$
$$^{-}1 \times {}^{-}1 = {}^{-}1$$
$$^{-}2 \times {}^{-}2 = {}^{-}4$$
$$^{-}3 \times {}^{-}3 = {}^{-}9$$
$$^{-}4 \times {}^{-}4 = {}^{-}16$$
$$^{-}5 \times {}^{-}5 = {}^{-}25$$

No explanation needed, and nothing in basic mathematics looks any more beautiful and as natural as that, it is how it was meant to be.

The New Formation of Balanced Calculations

Another way to see how the calculations now work in a very natural and even way:

$$1 + 1 = 2 \qquad 1 \times 1 = 1$$

$$^{-}1 + {}^{-}1 = {}^{-}2 \qquad {}^{-}1 \times {}^{-}1 = {}^{-}1$$

$$2 - 1 = 1 \qquad 2 \div 1 = 2$$

$$^{-}2 - {}^{-}1 = {}^{-}1 \qquad {}^{-}2 \div {}^{-}1 = {}^{-}2$$

This is how the calculation look to this day, the logic and sense of it is somewhat inane:

$$1 + 1 = 2 \qquad 1 \times 1 = 1$$

$$^{-}1 + {}^{-}1 = {}^{-}2 \qquad {}^{-}1 \times {}^{-}1 = 1$$

$$2 - 1 = 1 \qquad 2 \div 1 = 2$$

$$^{-}2 - {}^{-}1 = {}^{-}1 \qquad {}^{-}2 \div {}^{-}1 = 2$$

Distributive Property Now Works Better

Distribution with negatives could only be done traditionally if two negatives calculated to be positive; the reason for this is that the multiplicand was not indicated as being the focus of the problem at hand. When using the product of a negative as being negative, everything works as it should:

$$\begin{aligned} {}^{-}3({}^{-}4 + {}^{-}5) &= {}^{-}4 \times {}^{-}3 + {}^{-}5 \times {}^{-}3 \\ &= {}^{-}12 + {}^{-}15 \\ &= {}^{-}27 \end{aligned}$$

With the next example there are two multiplicands, one positive and the other negative, as well as the multiplier being negative, this still allows all the calculations to be correct:

$$\begin{aligned} {}^{-}3({}^{-}4 + 5) &= {}^{-}4 \times {}^{-}3 + 5 \times {}^{-}3 \\ &= {}^{-}12 + 15 \\ &= 3 \end{aligned}$$

The negative three outside the parentheses is the multiplier. Even if the two multiplicands are calculated first inside the parentheses (${}^{-}4 + 5 = 1$), the result stays positive because the multiplicand is positive. The multiplicand of 1 is being multiplied by the negative multiplier, resulting in 3.

Solving the Impossible Equation

Traditionally, the following equation is considered impossible to solve using only real numbers, because it is believed that squaring any value and then adding one can have no real-world solution. This stems from the axiom that no real number squared can yield a negative result:

$$(\boldsymbol{x})^2 + \mathbf{1} = \mathbf{0}$$

To bypass this perceived impossibility, mathematics created a workaround using imaginary numbers. Without going into exhaustive detail, the imaginary unit 'i' is assigned two complex states of being:

$$\boldsymbol{x} = \boldsymbol{i} \textbf{ and } \boldsymbol{x} = {}^{-}\boldsymbol{i}$$

This fix created a lasting complication: the radical symbol must almost always be preceded by a $\pm$ to allow mathematical concepts to function.

However, now that it has been discovered that the square root of a negative is negative:

$$\sqrt{{}^{-}1} = {}^{-}1$$

The logic becomes strikingly simple. Just as the positive territory is governed by:

$$(\boldsymbol{x})^2 + \mathbf{1} = \mathbf{0} \text{ (where } x = 1)$$

There is a logical necessity for an alternate, mirrored equation for the negative territory:

$$({}^{-}\boldsymbol{x})^2 + \mathbf{1} = \mathbf{0}$$

Because positive and negative are two individual and separate entities, this second equation is justified and exists in its own right as a functional mirror of the positive state. It is vital to note how this expression is read. Much like plotting points on a graph where an axis represents a distinct territory, the negative sign inside these parentheses is not an algebraic command to multiply or create a double negative. Instead, it explicitly defines the territory itself. When $\boldsymbol{x}$ is quantified as a negative value within this proper territory, the variable simply holds that negative entity directly. Substituting this entity into the equation yields a flawless physical balance:

$$(^{-}\mathbf{1})^{2} + \mathbf{1} = \mathbf{0}$$

Freedom from the distortions of standard double-negative rules reveals the true, elegant solution:

$$\boldsymbol{x} = {}^{-}\mathbf{1}.$$

10

Multiplication and its Concepts Explained

How does multiplication really work? There are two distinct concepts of multiplication used in mathematics. The first is simply calculating two or more parts together to ascertain the overall quantity. This is commonly known as basic multiplication. The multiplicand is the amount of objects or units, and the multiplier shows how many groups there are. They are both predetermined numbers within the mathematical story being read, and they are always present in one form or another. For example, if there are four baskets with six apples in each basket, the multiplicand and multiplier are fixed. Even if the mathematical story requires higher levels of math—such as calculating the distance a car has travelled—the multiplicand and the multiplier will be stated at some single point in time. If they appear in the capacity of a function, they are always present.

The other concept of multiplication involves a multiplier that can vary its value as needed, specifically when the outcome of the multiplicand is intended to be constantly changing to an unknown amount. The multiplicand in this kind of operation has a starting point and waits to see how much it will be increased or decreased; this is known as scaling.

A multiplier—whether it is positive or negative—cannot change the multiplicand from its original state of being (positive or negative). Whatever sign the multiplicand starts with is the sign it will retain at the end of the calculations; this is absolute. For example, two positive apples multiplied by a negative amount cannot make the apples negative.

The following three statements are true for any numerical value:

- When the multiplier is 1, the multiplicand remains the same amount.
- When the multiplier is greater than 1, the multiplicand increases.
- When the multiplier is less than 1, the multiplicand decreases.

These truths apply to all values: positive and negative natural numbers, whole numbers, integers, rational numbers, and even irrational numbers. It is now quite obvious that the multiplicand cannot reduce or increase the amount or size of the multiplier, as the multiplier is the tool used to manage the multiplicand.

In the previous chapter, the concept of the number line was corrected so that multiplication with negative values increases and decreases properly. Furthermore, this correction allows the concept of more is greater to be applied with a much better understanding of negative numbers: $^{-}7.4$ is greater than $^{-}7.1$. If the number line is not corrected, the following calculations do not work for all cases, as some will inevitably fail. Both multiplication concepts described above are used every day in a plenitude of natural ways throughout the world, and they always follow the structure shown below.

Examples of the Multiplier Being Less Than One:

$$0.5 \times 0.3 = 0.15$$

If the multiplier is 0.3 it is reducing the multiplicand, 0.5, to a smaller amount:

$$0.15 < 0.5$$

$$^{-}0.5 \times 0.3 = {}^{-}0.15$$

If the multiplier is 0.3 it is reducing the multiplicand, $^{-}0.5$, to a smaller amount:

$$^{-}0.15 < {}^{-}0.5$$

$$^{-}0.5 \times {}^{-}0.3 = {}^{-}0.15$$

If the multiplier is $^{-}0.3$ it is reducing the multiplicand, $^{-}0.5$, to a smaller amount:

$$^{-}0.15 < {}^{-}0.5$$

$$0.5 \times {}^{-}0.3 = 0.15$$

If the multiplier is $^{-}0.3$ it is reducing the multiplicand, 0.5, to a smaller amount:

$$0.15 < 0.5$$

Examples of the Multiplier Being Greater Than One:

$$0.5 \times 2.3 = 1.15$$

If the multiplier is 2.3 it is increasing the multiplicand, 0.5, to a greater amount:

$$1.15 > 0.5$$

$$^{-}0.5 \times 2.3 = {}^{-}1.15$$

If the multiplier is 2.3 it is increasing the multiplicand, $^{-}0.5$, to a greater amount:

$$^{-}1.15 > {}^{-}0.5$$

$$^{-}0.5 \times {}^{-}2.3 = {}^{-}1.15$$

If the multiplier is $^{-}2.3$ it is increasing the multiplicand, $^{-}0.5$, to a greater amount:

$$^{-}1.15 > {}^{-}0.5$$

$$0.5 \times {}^{-}2.3 = 1.15$$

If the multiplier is $^{-}2.3$ it is increasing the multiplicand, 0.5, to a greater amount:

$$1.15 > 0.5$$

Multiplication as Scaling

The second concept discussed above refers to what is known as scaling. This means there is an increase or decrease of an original quantity. The object or unit that the calculations begin with could be anything: a line, physical objects, digital data, money, time, distance, or speed. To modify the size of the original object when scaling, information is sought from the mathematical story. For instance, if one side of a rectangle is 5 centimetres long and it needs to be increased to a maximum of 50 centimetres, the sizes of the rectangle side must be determined for every 5 centimetre increment. The calculations are straightforward when using algebra to find the answers:

$$5 \times a = 50 \therefore a = 10$$

To modify the rectangle, all sides must be multiplied by the amount found, which is called the scale factor (in this case, the value of 10).

A simpler example involves a line with a set length of 4 centimetres. When scaling, the larger the multiplier is, the longer the line becomes; the smaller the multiplier is, the shorter the line becomes. For a line 6 centimetres long, this value is the multiplicand. Whatever values are used to increase or decrease the size of the line are the multipliers.

It is important to note that scaling does not have a special set of rules or functions set aside exclusively to work out associated problems; the basic fundamentals of multiplication are still used. Scaling is simply the application of those fundamentals to a changing quantity.

Multiplying by the Number Ten - *Do Not* Just Add a Zero, Ever.

When multiplying a number by ten, it is very well known that the decimal point does not move; in spite of this, it seems to occasionally find its way into some curriculum. Another very well-known idea is when a number is multiplied by ten, all the numbers move together to the left (not on a number line) and then a zero is added to give the answer. This concept is more widely adhered to than moving the decimal point. No matter which choice is taken to deal with this multiplying by a ten problem, both ideas fail at some point. When moving the decimal point, the failure occurs when multiplying a whole number, which indicates that the decimal point needs to move towards the right. This in turn causes the decimal point to be mysteriously sitting in a space that has no numbers either side of

it—just floating around. To make it logical, a zero must be added.
The second idea has the same failure as moving the decimal point. The moving of the number and then adding a zero has another problem, when multiplying a decimal such as 2.3561; the number can move, but the adding a zero now does not exist. For a number such as 2.3561, moving the decimal point or moving the entire number actually works perfectly well for both concepts; it is the adding a zero concept that is causing the problem. Why is it that a zero never needs to be added when multiplying a number by ten? The answer is quite simple: Those zeros are already there—just waiting for someone to multiply or divide. In mathematics, zeros are used in the most amazing places for totally awesome calculations, but some of the zeros that exist in different places are quite often forgotten. Zeros are always present before and after the decimal point, even if there are no other higher values. This is accepted and very much encouraged when a number is less than one (i.e., zero point six or 0.6), so it must be understood that the opposite is also correct (i.e., six point zero or 6.0). So, when a number is multiplied by ten, the number moves and the zero before the decimal point moves as well. As the zero moves into place—from directly before the decimal point to directly after the decimal point—it appears that a zero has just been added to the single number, hence the idea of adding a zero when multiplying by ten. It is very important to understand that a zero cannot magically appear out of thin air; it must come from somewhere. While invisible to all, they move with all other digits at all times.

In reality, a zero is never just added; it is always there, however as confidence grows with practice and, of course, greater knowledge, adding a zero correctly or moving the decimal point works very nicely. This concept is exactly the same when dividing by ten, but of course, the digits or the decimal points move to the right. Although this is not very important, the reality of what is happening is that the numbers do not move, the decimal point does not move—in fact, nothing moves at all. It is all just an illusion, well, of sorts anyway. When multiplying a number by ten, new numbers are formed for each place value; it just appears that the numbers are moving to the left. They are in fact increasing, not moving. But it is still alright to use the idea that the numbers are moving; it does make thinking about mathematics quite a bit simpler.

Is Multiplication Repeated Addition or Not?

Learning multiplication involves many steps, and one of those is, of course, repeated addition. The general process seems to be fairly unanimous throughout the world. As learners become more accomplished at performing multiplication, the idea of repeated addition becomes quite redundant—and rightly so. Teaching multiplication as repeated addition is rigorously debated among many who wonder if it will be detrimental to the way learners perceive mathematics as they move towards more advanced levels. Regardless of the debated outcomes, this book will use repeated addition—not because it is necessarily the best way, but because of how this book is going to reveal how all multiplication can be performed as repeated addition. From the moment a learner uses a memorised multiplication fact without relying on repeated addition to work out a problem, the method is no longer needed to be used nor thought about again.

Therefore, this section is not about how to teach repeated addition or how it should be learned.

The following multiplication as repeated addition examples are not shown here to be used in any practical sense. They are being shown simply because many sources state that representing all multiplication as repeated addition cannot be done and that immediately presented a challenge for the author. The remaining topics in this chapter show how multiplication works and recognises how repeated addition can be performed using that concept where needed.

Calculations of a Triangle

The problem of repeated addition with this example is that it contains a fraction, and that causes an aura of ambiguity. When visually taking in the possibility of calculating the area of a shape that does not contain all right angles, it does appear that repeated addition is not an option. To work out the area of a triangle, a simple formula is used: one half multiplied by the base and then multiplied by the height $(1/2 \times B \times H)$. This basic formula is commonly read as half the base by height. Generally, the base and the height are multiplied first, which leaves the multiplication of a half. The phrase multiplication of a half is the reason why performing repeated addition cannot be done to find the area of a triangle. After the base and the height are calculated, the multiplication of the half is done. This literally means: divide the product of the base and the height by two. The half is not really a fraction in this context; it is simply division.

$$\frac{1 \times B \times H}{2}$$

This, of course, is not anything new; it just helps to see the repeated addition. Performing multiplication of the base and the height is repeated addition, and the division of two is done separately and has nothing to do with repeated addition. This means that no, using repeated addition to find the area of a triangle cannot be done fully, but the multiplication part of the calculations can most certainly be done, and the rest is actually repeated subtraction.

How Multiplication of Two Never-Ending Decimals Can Be Done as Repeated Addition

Performing the multiplication of irrational numbers can be easily accomplished with the assistance of a data processor, such as a supercomputer, or a more compact device like a handheld calculator. It is believed that these irrational numbers consist of millions of digits; therefore, when performing manual calculations, particularly with multiplication, these irrationals need to be approximated to a specific number of decimal places. It is advantageous to round a large decimal to a reasonable number of places in order to facilitate more efficient problem-solving. The processes of multiplying two numbers—for example, pi (π) and the square root of two ($\sqrt{2}$) by repeatedly adding them together demonstrates the same rounding principles used in standard multiplication:

$$314{,}159{,}265{,}359 \times 141{,}421{,}356{,}237 = \text{a very large number!}$$

Utilising multiplication as repeated addition to solve the aforementioned number sentence would be an incredibly inefficient task. It would require a significant amount of time and a substantial amount of paper, but it is not impossible. The process of repeatedly adding these two numbers would involve creating $141{,}421{,}356{,}237$ groups, with each individual group consisting of the number $314{,}159{,}265{,}359$, and then adding every number from every group together. In other words $314{,}159{,}265{,}359$ would be written down $141{,}421{,}356{,}237$ times and then added together. This exercise was just to show how significant rounding will be needed to be able to multiply pi and the square root of two by hand. Rounding to this many place values will be similar to the above example:

$$3.14159265359 \times 1.41421356237$$

Reducing the number of digits even more would undoubtedly simplify the task:

$$3.14159 \times 1.4142$$

Prior to demonstrating the process of repeated addition as a method for multiplication for these two numbers, it is essential to first explain how a single-digit number multiplied by a single-digit decimal works. Utilising groups as a tangible method for teaching multiplication provides a valuable opportunity to demonstrate the concept of repeated addition. Regrettably, in current mathematics there is a lack of clear explanation for applying decimals as groups when used as a multiplier; nevertheless, the fundamental concept remains consistent with that of whole numbers. This section provides a clear method to accomplish this task in a simple manner. While this example may not be the only approach, it effectively demonstrates how decimal multiplication can be viewed as repeated addition. To initiate the process, it is crucial to establish a fundamental framework by ensuring that it is recognised as a proven and flawless method. It is important to note that throughout this chapter, the multiplicand will continue to be designated as the first number and the multiplier second.

$$3 \times 4$$

This expression is very simply stated as three multiplied by four. To see it as repeated addition, it could be said as four groups with three in each group. This statement will serve as the foundation for further testing of each expression. Repeated addition for this can simply be seen as $3 + 3 + 3 + 3$, which sums to 12.

The next example is an idea that is already well known, so consistency will be shown:

$$0.4 \times 2.0$$

This is very simple to read as well: two groups with zero point four in each group. Repeated addition for this is standard and can be seen as $0.4 + 0.4$, which sums to 0.8. This is the point at which the presentation of repeated addition in decimal multiplication concludes for most learners, as commutation is usually used to calculate the following problem that has the multiplier as a decimal:

$$3 \times 0.2$$

This is the first expression that has a decimal as the multiplier (0.2). To be consistent, the sentence reads as: zero point two groups with three in each group. While this demonstrates consistency, it appears that this concept lacks coherence, as it seems implausible to have zero point two groups. However, the following will show how this is possible. When using a whole number such as 2 to show repeated addition, two groups are formed with each group having a value of 1. This approach can be used for the multiplier of 0.2. Ten groups are formed, such as a circle divided into ten groups (tenths) but only two of those groups will be used, and each of those two groups has a value of 0.1. The hard part to see is that the original whole number multiplicand cannot appear in the decimal groups, if one whole unit was placed in each decimal group then there would still be 2 groups with 1 unit in each. The multiplier is instructing the multiplicand to come in line with the value of the multiplier. Because the multiplier is in the form of tenths, it is instructing the multiplicand to become the same place value—dividing it by ten. Subsequently, that value will be placed into each group. If the multiplier were hundredths 0.02, the multiplicand would be divided by one hundred.

- 3×0.04: There will be 4 groups of 0.01—that is 0.03 in each group.
- 5×0.0029: There will be 29 groups of 0.0001—that is 0.0005 in each group.
- 628×0.26: There will be 26 groups of 0.01 —that is 6.28 in each group.

For a decimal multiplier greater than one i.e. 1.5 the process is exactly the same but with a bit more work to do. To achieve this, it must be understood that there cannot be 1 group having a whole number value of 1 and 5 groups having a decimal value of 0.1, they both must be of the same place value to be able to be calculated in a repetitive manner. So the 1 of the multiplier will be changed into the same value as the decimal part of the multiplier. The decimal part is tenths, for the 1 there will be ten tenths giving a total of 15 groups of 0.1:

- 4×1.5 there will be 15 groups of 0.1 - that is 0.4 in each group of 0.1

What if there were a decimal multiplied by a decimal? The process is the same:

- 6.3×1.7 there will be 17 groups of 0.1 - that is 0.63 in each group of 0.1

The following expression is calculated by utilising the base model outlined in the multiplication section of Chapter 7:- 3×0.6
The multiplicand is $^{+}\mathbf{3}$ and the multiplier is $^{+}\mathbf{0.6}$
The sign of the multiplicand states which side of the number line that the calculations will take place, which is the positive side. The multiplication operator states which way to face on the predetermined side of the number line, which is towards positive infinity. The multiplicand states how far to move on the number line per jump and this is $\mathbf{0.3}$ positive units per jump, $\mathbf{0.1}$ step at a time. The multiplier states how many times the multiplicand moves repeatedly on the number line or how many jumps will be made, which is $\mathbf{0.6}$ jumps. The sign of the multiplier states what environment the calculations are taking place, which is in a positive environment.

$$\textbf{Hence: } {}^{+}\mathbf{3} \times {}^{+}\mathbf{0.6} = {}^{+}\mathbf{1.8}$$

Finally, here is the solution to repeatedly adding two never-ending decimals:

$$3.14159 \times 1.4142$$

To show repeated addition, the 1 in the multiplier of 1.4142 will be divided into 10,000 divisions (because the decimal part is in the ten-thousandths place). Now there will be 14,142 groups with the value of 0.0001 each. The multiplicand is divided by ten thousand to become 0.00031459. To repeatedly add 0.00031459, the value would be written down $14{,}142$ times and then summed together:

$$14{,}142 \text{ groups of } 0.00031459 = 4.44893178$$

While some might argue that showing 3×0.04 as four groups of 0.03 is simply the equivalent of dividing 3 by 100 and multiplying 0.04 by 100 to get 0.03×4, that misses the entire point. Standard mathematics uses that manipulation as a superficial shortcut. This topic demonstrates the underlying reality: how to physically structure repeated addition, proving exactly why dividing and multiplying to restructure the groups actually works in physical space.

Repeated addition using a fraction as a multiplier is a physical impossibility because fractions represent parts, not whole countable groups. For instance, attempting to calculate $3 \times 1/2$ as repeated addition fails immediately; you cannot execute half a group of addition. To use a fraction as a multiplier, the operation must be recognised for what it actually is: a division problem disguised as multiplication

Repeated Addition of Scaling Objects

When enlarging or shrinking an object such as square or a circle, a rubber band (a slightly different concept of stretching where there is no amount of the rubber band lost or gained) or an image on a personal electronic device would mean to calculate every movement that was made. Using a line six centimetres long as an example, it increases with every new multiplier applied so therefore a new calculation is performed each change in size (lines not to proportion):

6 cm, or

6×1 ____
6×2 ______
6×3 ________
6×4__________

The exact same process is performed to make the same line shorter:

6 cm, or

6×1 ____
6×0.5 __
6×0.4 _
6×0.3 _

This of course is nothing new and is easy to see that numbers greater than one can be done as repeated addition. The numbers below one is now performed easily as shown in the previous topic. The difficulty here is the amount of new sizes that need to be calculated. This includes all the decimals in-between three and four, which would entail a substantial number of calculations. This expansion or reduction process is better understood if it were seen as being calculated as an innumerable amount of stages. If these calculations were done by hand, the physical movement of the object would need to have a single calculation done with each adjustment made. This would of course be nearly impossible because the time frame to calculate every interval would be in the range of infinity. Although, computers have certain algorithms to accommodate any size required. If restraints were placed in order to do single calculations at small intervals, then scaling could be done easily using normal multiplication as repeated addition at every one of those chosen intervals.

Multiplying Decimals - Finding the Placement of the Decimal Point Mathematically

When performing decimal multiplication, it is widely known that to achieve a correct answer, the process first requires ignoring the decimal points in the numbers being multiplied together. After completing the necessary multiplication, the places of the decimal points in both of the original numbers are counted and then combined. The next step is to take that amount and, counting from the right to the left of the product, place the decimal point in the counted place. This of course is very effective, and also very fast. The proper method of working out the correct decimal place in the product, displayed below, is quite time-consuming—especially when learning how to do this new way. But this is not about how to teach this concept; it is about understanding how the process is actually working when column multiplication is performed.

$$2.94 \times 0.6 =$$

	0.04
2.94	$\times\, 0.6$
$\underline{\times\, 0.6}$	This is a hundredth multiplied by a tenth so the
$\_.\_\ \_\ 4$	answer will be in the thousandths.
	$\_.\_\ \_\ \_$
	$4 \times 6 = 24$ so 24 thousandths or 0.024
	Place the 0.004 below the vinculum and carry the 0.02 to be placed above the $.9$

$$\begin{array}{r} 0.9 \\ \times\ 0.6 \end{array}$$

$$\begin{array}{r} {\scriptstyle 0.02} \\ 2.94 \\ \times\ 0.6 \\ \_.\_\ 64 \end{array}$$

This is a tenth multiplied by a tenth so the answer will be in the hundredths.

_._ _

$9 \times 6 = 54$ so 54 hundredths or 0.54

Add the carried over amount of 0.02 to 0.54 which equals 0.56

Place the 0.06 below the vinculum and carry 0.5 to be placed above the 2.

$$\begin{array}{r} 2 \\ \times\ 0.6 \end{array}$$

$$\begin{array}{r} {\scriptstyle 0.5} \\ 2.94 \\ \underline{\times\ 0.6} \\ 1.764 \end{array}$$

This is a unit multiplied by a tenth so the answer will be in the tenths.

_._

$2 \times 6 = 12$ so 12 tenths or 1.2

Add the carried over amount of 0.5 to 1.2 which equals 1.7

Place the 0.7 below the line in the tenths place and the 1 in the units place and that will give the final result.

In effect, the placement of the decimal point is still positioned three spaces from the right-hand side, but this is a mathematical calculation. The original method is simply basic counting and, at best, is a guess which does have a potential for making mistakes, such as misplacing the decimal point. This method forces the decimal point into the correct position, especially if there are trailing zeros.

No Units the Same: Distance = Speed × Time.

This example appears to be too complex to be able to perform repeated addition simply because there are three different types of units presented in the statement.

$$\frac{D}{S} = T$$

$$\frac{D}{T} = S$$

$$T \times S = D$$

A mathematical story for this could be how far did a person travel in two hours when walking at an average speed of four kilometres per hour? To be able to understand this more clearly, defining each unit will be necessary in order to work out the distance. Adding extra information will heighten the chances of being able to identify each piece of information more accurately:

Distance in kilometres = Speed in kilometres per hour × Time in hours

The distance will be the product, which is in kilometres. Crucially, the product must come from the same unit type as the multiplicand. Since time is measured in hours and is not a measure of distance, it must be the multiplier. This leaves speed, which is measured in kilometres (albeit kilometres per hour) to be the multiplicand.

With that information, it is easy to see multiplication as repeated addition at work:

Multiplicand × Multiplier = Product

Speed × Time = Distance

4 Km/h × 2 hours = 8 Kilometres

$$4 \times 2 = 8$$

$$4 + 4 = 8$$

If distance is the subject of the inquiry, it cannot be the multiplier or the multiplicand; it is always the product. This means that whatever unit is used for the final distance, the multiplicand must carry that same spatial unit. The multiplier cannot change the fundamental nature of the multiplicand. In a distance inquiry, time is measured in hours or minutes, not kilometres. Speed, however, is measured in kilometres (per unit of time). Therefore, speed must be the multiplicand—the subject being repeated—while time acts as the multiplier instructing how many times that speed is accumulated. Because the multiplier cannot change the kilometre nature of the speed, the resulting product is purely distance.

Multiplication is Not in Any Way Repeated Subtraction

A problem discovered in a maths discussion forum many years ago suggested that multiplication is also repeated subtraction. The post suggested:

$$2 \times 3 = 2(0 + 1 + 1 + 1) = 6 \text{ (Repeated Addition)}$$

$$2 \times -3 = 2(0 - 1 - 1 - 1) = -6 \text{ (Repeated Subtraction)}$$

Although this looks perfectly correct, the execution is somewhat misguided. There are two issues that need to be rectified.

The First Issue: Confusing Signs with Operators.

Positive and negative signs are being used as addition and subtraction operators. Focusing on the -3, it is clearly being presented as a negative, but within the parentheses they are clearly shown as 3 being subtracted, which of course is not the same thing. This is because each 1 inside the parentheses is positive and for those to equal negative three, it would need to add negative one three times:

$$^{+}2 \times {}^{+}3 = {}^{+}2(0 + {}^{+}1 + {}^{+}1 + {}^{+}1) = {}^{+}6 \text{ (Repeated addition of positive values)}$$

$$^{+}2 \times {}^{-}3 = {}^{+}2(0 + {}^{-}1 + {}^{-}1 + {}^{-}1) = {}^{-}6 \text{ (Repeated addition of negative values)}$$

Both number sentences are written with the multiplier first and the multiplicand second. When setting up the distribution, the multipliers are on the outside of the parentheses.

The Second Issue: Timing of the Calculation

The actual calculations considered to be repeated addition and subtraction are performed at the wrong time in the original example. When multiplication is done during distribution for the original example, it is shown within the parentheses; this is the moment that the supposed repeated subtraction is performed. The -1 inside the parentheses are of course the result of the deconstruction of 3 and have nothing to do with multiplication. If the veil is lifted from the calculations to show exactly where the repeated addition is occurring, then there is absolutely no doubt that multiplication is repeated addition for both positives and negatives. The following two examples have the multiplicand first and the multiplier second with the multiplier outside of the parentheses:

$$\begin{aligned} {}^{+}4 \times {}^{+}3 &= {}^{+}3(0 + {}^{+}2 + {}^{+}2) \\ &= 0 \times {}^{+}3 + {}^{+}2 \times {}^{+}3 + {}^{+}2 \times {}^{+}3 \\ &= (0 + 0 + 0) + (0 + {}^{+}2 + {}^{+}2 + {}^{+}2) + (0 + {}^{+}2 + {}^{+}2 \\ &\quad + {}^{+}2) \end{aligned}$$

This is repeated addition of positive two.

$$
\begin{aligned}
{}^{-}4 \times {}^{+}3 &= {}^{+}3(0 + {}^{-}2 + {}^{-}2) \\
&= 0 \times {}^{+}3 + {}^{-}2 \times {}^{+}3 + {}^{-}2 \times {}^{+}3 \\
&= (0 + 0 + 0) + (0 + {}^{-}2 + {}^{-}2 + {}^{-}2) + (0 + {}^{-}2 + {}^{-}2 \\
&\qquad + {}^{-}2)
\end{aligned}
$$

This is repeated addition of negative two.

11

Collatz Conjecture a Negative Conundrum

This section is not about solving this wonderfully infamous maths problem which in itself is considered undecidable. However, there is another issue with this conjecture, and that is the negative aspect when testing negative values. When performing the calculations of $3n + 1$ and the division of 2 on the positive side of the number line, all numbers (well, actually the first quintillion numbers that have been tested) come to what is called a $4 \rightarrow 2 \rightarrow 1$ loop. The main problem is that this $4 \rightarrow 2 \rightarrow 1$ loop does not exclusively occur on the negative side of the number line; in fact, there are several different loops. As the concept of negative is an opposite of the concept of positive in mathematics, there must be a reason why the negative calculations are not giving the exact same opposite results. Finding a solution for why this is occurring will involve using information that has previously been discussed in this book.

The primary concept to consider is to look at the parts of the function $3n + 1$ as well as understanding the division by two is simple and does need analysing. The function $3n + 1$ has two operators in use: multiplication and addition. The multiplicand, as always, is the main focus of each individual calculation and will stay that sign (positive or negative) regardless of the sign of the multiplier. The multiplicand in this function is represented by

the variable n, which is the only element that undergoes continuous change with each new number tested. There is also a positive addend, the number 1, which is added to the product of $3n$, and the last value is the 3, which is the multiplier and is also positive. It must be remembered that multipliers and divisors cannot change the value or the sign of the multiplicand.
When inserting a positive number (e.g. 7) into the *n* part of the function $3n + 1$, the resulting expression looks like this:

$$^{+}3 \times {}^{+}7 + {}^{+}1$$

If a negative number is used (e.g. $^{-}7$), it will be expressed like so:

$$^{+}3 \times {}^{-}7 + {}^{+}1$$

Comparing the two expressions, it is clear to see that only one thing has changed. For these calculations to work for negative values, the formula must be the opposite of the function that is on the positive side of the number line.
Testing $^{+}5$ against $^{-}5$ shows a different outcome and a different loop:

- Positive: $5 \rightarrow 16 \rightarrow 8 \rightarrow 4 \rightarrow 2 \rightarrow 1 \rightarrow 4 \ldots$
- Negative: $^{-}5 \rightarrow {}^{-}14 \rightarrow {}^{-}7 \rightarrow {}^{-}20 \rightarrow {}^{-}10 \rightarrow {}^{-}5$ ... (the pattern repeats)

The first calculation of each is:

$$3 \times 5 = 15 \text{ and then } 15 + 1 = 16$$

$$3 \times {}^{-}5 = {}^{-}15 \text{ and then } {}^{-}15 + 1 = {}^{-}14$$

When studying these two sets, it is obvious that when multiplying the input of $^{+}5$ and $^{-}5$ by the constant of 3, the results are exactly the opposite of each other. However, when each product has one added to it, it becomes evident that this is the point where the calculations stop being consistently opposite. As the calculations for the positive results move towards positive infinity, the negative results should be completely opposite and move towards negative infinity—but they clearly are not.

If the addition of one were changed to subtraction, it would look like this:

$$^{+}3 \times {}^{-}5 - {}^{+}1$$

That would see the calculations working perfectly for all negative numbers

$$^{+}3 \times {}^{-}5 = {}^{-}15 \rightarrow {}^{-}15 - {}^{+}1 = {}^{-}16$$

However, the opposite of $3n + 1$ is not simply the subtraction of 1. When the multiplicand is negative, changing the addition of 1 to the subtraction of 1 causes two concerns. First, changing the function from addition to subtraction would mean an entirely new function has been created. Second, the 1 being subtracted is still a positive, and as established, a positive cannot exist within a negative space. Changing the addition of a positive number to the addition of a negative number solves this.

In the original formula, $3n + 1$, the addition of 1 is instructing the calculations to add one more. Where does the 1 come from? The 1 is more of n, meaning that 1 is the same entity as *n*. On the logical inference that negatives are the opposite of positives, it stands to reason that if the *n* of the function $3n + 1$ is used on the negative side, and then the whole function must be considered opposite as well. To maintain consistency within the new framework, the operation must reflect the opposite identity of the original function:

$$3n + 1$$

This, in the negative domain, is defined as:

$$^{-}3 \times {}^{-}n + {}^{-}1$$

This still recognises that the addition of $^{-}1$ is the opposite of the addition of $^{+}1$ and is a valid part of the conjecture because it is still adding.
The reason why $^{-}3 \times {}^{-}n + {}^{-}1$ is not an inverse, is simply because it does not undo what $3n + 1$ does. This process works beautifully under the new calculations presented in this book. Yet, even in standard modern arithmetic, it operates without requiring a negative multiplier, simply because a multiplier has no effect on the underlying sign of the multiplicand. So, the new formula for the negative side of the number line is:

$$3 \times {}^{-}n + {}^{-}1$$

And that is how to fix the Collatz conjecture for negative numbers.

12

How Base Ten Works and Why it is Perfect

The most basic function of mathematics is, of course, counting, and the way it is currently done appears to be in many sized bases the world over. For the majority, though, the most common and widely used base is that of ten. It is base ten simply because it has ten numbers, which are zero to nine. Base ten is a repeating growth numbering system and is what makes place value, in any base, possible. This growth system is quite amazing in how something so basic works so effortlessly, and it is possibly the greatest achievement in the history of the world of mathematics. This is all possible because of one number, and this is the number zero.

A repeating number system must have all numbers repeating such that at the end of each cycle of the pattern, all of its numbers appear again and again. When counting, as seen on paper, the base ten system must start from zero and end with nine, directly followed by a zero and so on, which of course is something that is well known. The next requirement is being able to use those numbers to keep recycling the nine and then to the next zero to produce higher numbers, but how is this done? It is not common practice but still known that in front, written to the left of a number, there is an infinite amount of insignificant zeros.

These are used to allow the counting system of base ten to grow by counting up from zero, or else there is nothing to count up from and the growth function could not work:

00
01
02
⋮
08
09
10
11

And so on.

When nine is reached in the units place value, the pattern starts again from zero; this is just one amazing aspect of zero where it can be seen as a very special number. When **0** appears at the start of a new cycle, it instructs the previous **9** to command the number to its left, in this case **0**, to go up by one to **1**. Zero is the only number that performs this unique function. This concept shows how counting to ten is not just some random event; counting up is able to go from nine to ten because it is a natural process done in a very methodical manner:

189
190

This example explains that when the pattern reaches **0** for the units place value, the previous **9** commands the **8**, in the tens place value, to count up by one, to **9**. When **9** starts in the new pattern, it will not count up to the next number of zero until a zero starts the command process again in the units place value, and thus will continue producing nines until then:

$$
\begin{gathered}
191 \\
192 \\
\vdots \\
197 \\
198 \\
\mathbf{199} \\
\mathbf{200}
\end{gathered}
$$

When $\mathbf{0}$ in the units place value is reached again it commands the previous $\mathbf{9}$ to command the $\mathbf{9}$ next to it, to go up by one to $\mathbf{0}$. When this $\mathbf{0}$ is reached in the tens place value, that zero commands the previous $\mathbf{9}$ to command the $\mathbf{1}$ in the hundreds place value, to the left of that nine, to count up from $\mathbf{1}$ to $\mathbf{2}$. As this process continues on forever, it makes the base ten numbering system unique due to the fact that no other base counting system has this grandiose quality. This in turn is what makes the base ten counting system perfect. This system is not perfect because humans have ten fingers; although it most certainly may have been created for that very reason, it is perfect because of its simplicity and the amazingly clever repetitive growth function. Before continuing, a quick glance at the most common bases, not including computers, in use today:

Base $\mathbf{10}$ $(\mathbf{0}$ to $\mathbf{9})$: $\{0, 1, 2, 3, 4, 5, 6, 7, 8, 9\}$

Base $\mathbf{12}$ $(\mathbf{0}$ to $\mathbf{11})$: $\{0, 1, 2, 3, 4, 5, 6, 7, 8, 9, 10, 11\}$

Base $\mathbf{16}$ $(\mathbf{0}$ to $\mathbf{15})$:

$$\{0, 1, 2, 3, 4, 5, 6, 7, 8, 9, 10, 11, 12, 13, 14, 15\}$$

There are two main systems that are currently used throughout the world, and they are the metric and imperial systems. The metric system is very well known, which does not at any time faulter and always uses base ten for measurements and weights, measures, and money, while money is not a metric unit, it still uses base ten. Countries that use the imperial system have a variety of different bases, such as base twelve for measurement and base sixteen for weight, but use the easier system of base ten for money and all electricity. In fact, every single electrical device, large and small, on the face of the entire planet uses the metric system as the base unit of electrical measurement.

The Dozenal System and Other Systems That Replicate Base 10 Do Not Follow the Laws of the Natural World

For imperial measurement, the term base twelve is not quite accurate, such that there may be multiplies of twelve from one level of measurement to the next level. This is evident when seeing twelve inches to a foot or three feet to one yard, all of which are multiplies of twelve (e.g. thirty-six inches to one yard). But as distances grow larger, the base of twelve disappears (e.g. one furlong is two hundred and twenty yards, which is not evenly divisible by twelve). Hence the reason why the imperial system for measurement is not base twelve.

On an imperial ruler or measuring tape, decimals are never seen; fractions of an inch or other fractions are always present. Once upon a time, measurements were always done as fractions, but at some point in time, they changed to being able to use decimal inches when needed, which may have been about the 19th century. This is still evident today (i.e., when cutting a round pizza into four pieces it results in 0.25 per piece). In this day and age, decimal inches are clearly used for the size of electronic devices as well; for example, a 10.1 screen of a tablet is an imperial measurement in decimal inches.

If a pizza were cut into three pieces, it is equal to a repeating decimal, which of course is 0.33333... This is why fractions are used to show an easier number to comprehend: 1/3. Another example of a concept that uses different bases is time. Clocks or some form of time-keeping devices are either digital or analogue, comprising either twelve or twenty-four hours in a day. When presented in an analogue way, the numbers appear as one to twelve usually, but this is for human simplicity which is still considered base twelve. Time is actually base twenty-four for hours and base sixty for minutes and seconds, but still was evolved from base twelve. These two number bases, sixty and twenty-four, never appear in their repeating patterns for digital output but can appear on an analogue clock face. Time also has another base which is base ten, and this is in the form of decimals and can be found for amounts less than a second, which is a millisecond. Before the adoption of the millisecond, time less than a second was measured in fractions; this, of course, changed at some point to be known as it is today. Any number counting base that uses the Hindu-Arabic numbering system is using base ten to create those bases.

There is a new counting system which is about five hundred years old, and that is the dozenal system. It uses two letters to represent the numbers ten and eleven, and the twelfth number is ten. This system works equally as well as base ten, but the only reason why this is so is because it uses the incredibly powerful attributes of the number ten that can be found only in the base ten system.

While the dozenal system appears to be perfect, it fails in one of the most critical areas, and is an area for which the dozenal system prides itself on fixing or making life much easier. This area that fails is of division by three. Calculations for the dozenal system have one divided by three as being 0.4 per piece. Now everyone knows that a circle cut into three pieces can never have even pieces. It does not matter what the numbers are called—they could be letters, lines, glyphs, or whatever. The simple fact remains that in a physical world—and all mathematics is real world—a circle that is cut into three pieces using division can never have exactly the same amount in each piece. The dozenal system claims to simplify division by three, yet it ignores the physical reality that a circle divided into three can never be perfectly equalised by a change in numbering. It is an attempt to use the power of the base ten system to justify a system the world has already outgrown.

This area only fails in a small way, but as discussed in previous chapters, if one concept fails even at the smallest parts, then there is something wrong, which means the whole concept will eventually fail. A circle may be cut into three even pieces and considered even (0.4) if they are fractions, which were discussed at length in Chapter 4.

The dozenal system could possibly benefit those countries that use base twelve and have a need to add fractions for their measurements. But for the rest of the world, people who use the base ten system may never need to add fractions for their entire lives and never need to learn the eleven and twelve times tables because there is no need to divide anything by twelve or to divide twelve into anything. Fractions and the twelve times table are utterly useless in a base ten/metric world and this is another reason why the base ten system is so perfect. After metrification of the entire world takes place—which it most certainly will—base 12 and the dozenal system will never need to be mentioned again, ever!

13

New Possibilities

Graphs of Linear Equations

When it comes to multiplication of both negative and positive numbers for plotting values on a graph, the multiplicand and the multiplier still exist. Finding out which is the multiplicand is somewhat challenging but very important to discover in order to see if there is now a change in the way graphs look. To find the multiplicand, studying how the graphs work and what the numbers mean will shed light on not just the multiplicand, but also how two negatives contribute to the creation of the graphs from the equations given.

The foundational model to examine is a linear equation which forms the simplest graph, and that is $y =$ any number. For this example, the equation will be:

$$y = 2$$

The y in this equation is simply the dependent variable, which literally depends on the outcome of calculations from the RHS (right-hand side) of the equation, or what x or other variables equate to. Generally speaking, y itself is not a multiplicand or a multiplier; and is thought of as being positive at all times. The y is what shapes the graph, and even though x and y work together to create it, it is the y that ultimately gives the answer to the question being asked, but more on that shortly. The value of 2 in this equation is what is known as a constant. Because it is by itself, it could possibly imply that it is an augend or an addend; it could also possibly be considered a multiplicand or a multiplier as well, but these ideas are unclear at this point in time. This equation does not seem to contain an x or any variable on the right-hand side, but as there are many shortcuts in mathematics with some values simply being implied as being in a certain place, it must be considered that a variable may be present. Without too much effort in discovering an answer, a very well-known and simple solution comes in the form of a line equation:

$$y = mx + c$$

To form the equation $y = 2$, the m would need to be equal to zero, which in turn makes the multiplication of m and x equal to zero, leaving c by itself—which, of course, equals 2. This brings the equation back to $y = 2$, which is wonderful because it means that 2 is, in fact, an addend and has nothing to do with the multiplication process. The main focus now moves to the m and the x of a slightly more complex linear equation.
The next example:

$$y = x$$

This is quite different from the first example because of the change from a constant number to a variable. The number in the first example does not change and is constant (always the y-intercept), whereas the x represents a change in any value that can be substituted to allow for more information to be placed onto the graph. To find out how the x is used with each new value, the way it is calculated needs to be examined. The most common understanding of the way x is perceived, after solving for it, is to say: "When x equals the inserted value, y equals that final calculated amount." Those calculations are made from equations that contain x, such as

$$y = x + 3$$

Equations like these have obvious calculations to make, but when there is an equation such as $y = x$, there does not seem to be any way to perform calculations for x. It appears to be just like the issue found with the previous equation of $y = 2$. If the equation for a line is used once again, a simple solution becomes apparent:

$$y = mx + c$$

The equation $y = x$ has an invisible shortcut, which is m, and in this case, it is a 1. The c is the y-intercept, and for this equation, it is zero. This allows a way to process calculations correctly without just substituting numerical values into x:

$$y = 1x + 0$$

This, of course, is nothing new, but it is important to understand that calculations must always occur—even if they are being done behind the scenes—in order to be mathematically consistent throughout. This is especially true for equations used for graphs, as they will always encompass multiplication while exhibiting a multiplicand and a multiplier.

Just like any number sentence, an equation for a graph will come from a mathematical story. It will provide all the necessary information about the problem in question, which means that the graph is created *after* the equation is written. Once the graph exists, substituting values into the equation via the variable (usually x) will indicate predetermined points on the line. These calculations have no effect on the shape of the graph itself; the moment the equation is created, the graph has its form. Inserting different values for x simply finds specific points of y which are the answers to the original question. With this knowledge, it is easy to see that x is a multiplier, as all other values are set and do not change.

The multiplicand is the coefficient m because it determines the gradient of the line—and the gradient is what the question is all about. This coefficient can be negative or positive. If there is no visible coefficient in front of the variable x, there is still a number: an invisible 1 which holds a positive or negative value.

This raises a vital issue regarding the value of x. Since there is always a numerical value between the positive or negative sign and the x itself, is the x positive, negative, or possibly... does it have no sign value at all?
To find the answer, the function of x and the way it is used during calculations must be closely examined. The primary concept to look at is how negative values, such as in the equation $y = {}^{-}x$, are calculated. To have a clear understanding of the negative values, positive values will be examined first. When creating a table of values for the equation:

$$y = x$$

The following setup is adopted:

x	y
$^{-}2$	
$^{-}1$	
0	
1	
2	

When generating values on a new table for the equation:

$$y = {}^{-}x$$

that table is always set up exactly the same way as above and has no negative sign connected to the x:

x	y
$^{-}2$	
$^{-}1$	
0	
1	
2	

When x is shown to be negative in this equation but not present in the table, the substituted values for x are made in the same way as the first example. To show negatives and positives more clearly, superscript signs will be used. Substituting x for $^{-}2$ as it is presented in the tables above will help to see how x is working and what its function is. Note that the following calculations are derived from the rules of the new framework:

$y = {}^{+}x$ When $x = {}^{-}2$ $y = {}^{+}\mathbf{1}(x)$ $y = {}^{+}1({}^{-}2)$ $y = {}^{+}2$	$y = {}^{-}x$ When $x = {}^{-}2$ $y = {}^{-}\mathbf{1}(x)$ $y = {}^{-}1({}^{-}2)$ $y = {}^{-}2$

This, of course, is nothing new in standard practice, but it clearly shows that x is neither positive nor negative and strongly suggests that it is merely a placeholder. This neutral status ensures that whatever value is substituted into x can freely function as a multiplier, fully retaining its positive or negative identity.

Zero is neither positive nor negative and can be substituted into x, but x is never zero while it is a placeholder. If it were, all calculations for multiplication would be zero at all times, and that obviously cannot happen.

Now that the roles of the multiplicand and the multiplier have been clearly established, graphs can be constructed and fully understood. By prioritising the most critical concept—the identity of the multiplicand—negative and positive values for graphs of all shapes and sizes can now be mapped with absolute accuracy.

Graphs of Circles

The general equation for the formula of a circle follows below:

$$(x - h)^2 + (y - k)^2 = r^2$$

It is used in a large variety of applications in real life and is also used to graph circles on a Cartesian plane for educational purposes. When finding points on the graph, there are always two coordinates for every x value, except for the outer most turning points on either side of the circle. To find these sets of points, the second-to-last lines of the calculations will have a plus and minus sign added to enable two different values for y. These two signs are exclusively addition and subtraction, as negative values will not work. This is indicated with a combined sign before the radical, $\pm\sqrt{}$. The following demonstrates a foundational way of observing how the calculations for a circle are constructed. The addition operator calculations create the top half of the circle, and the subtraction operator calculations create the bottom half.

When solving for y and the radius squared on the (RHS) of the equation, it is important to understand that no matter the numerical amount substituted into x, the resulting value within the first parentheses must always be positive.

$$(x-3)^2+(y-3)^2=2^2$$

The term highlighted above is the full value that will be calculated for x. Substituting a value into x must always be within the limits of the radius. If the total value for x component resulted in a negative value, it would end up being added to the right-hand side during calculations, making the calculated radius larger than the physical circle—which can never work properly. This is why the result must be positive so it can be subtracted from the RHS towards zero.

The x value can be interpreted in two different ways. For example, when a smaller value is substituted into x, the value will result in a negative number:

When $x = 2$:

$$(2 - 3)^2 = ({}^{-}1)^2$$

In standard mathematics, the system asserts that $({}^{-}1)^2$ must be equal to positive 1, thus subtracting from the RHS. However, the problem that has now arisen is that this book has stated that the product of two negatives is negative. There must be another way to perform this calculation. When looking at the above calculation, the following expression cannot equal a negative output, as that would break the physical bounds of the radius on the Cartesian plane:

$$(2 - 3)^2$$

Therefore, it can only represent one reality: a spatial difference, or a change in x. The two values in the parentheses are positive and are not negative at any time; therefore, the difference between two and three is one. This concept works with whole numbers and decimals:

When $x = 2$:

$$(2 - 3)^2 = (1)^2$$

Understanding this distinction is vital because when negative values for the radius are utilised, the calculations operate in exactly the same manner, but in the opposite direction.

The next phase of the circle framework explains how the values for the bottom half of the circle are obtained. With subtraction being the only logical possibility—solely because the product of two negatives is negative—negative directional values can now be introduced. Before continuing, a firm grounding in the components of the Pythagorean Theorem is required, beginning with the hypotenuse, which represents the radius. To incorporate negative values, a precise application shift must be made to the current equation. It is highly probable that if standard mathematics had originally recognised that the product of two negatives is negative, a completely different structural equation would have been established from the beginning.

The mechanism by which this equation functions is highly efficient to implement, though it requires a shift in perspective to grasp initially. In classical physics, the trajectory of an object thrown from a height is assessed by designating upward motion as positive and downward motion as negative. To distinguish between the two distinct points deriving from a single x value, this exact directional concept is applied to the graph of a circle. Positive values dictate movement towards positive infinity (upward on the Cartesian plane), while negative values dictate the opposite direction (downward towards negative infinity). More precisely, this spatial direction stems directly from the centre of the circle. Above the centre axis is positive; below it is negative. Consequently, the radius can carry a negative designation to explicitly indicate a downward direction on the plane. When studying the permissible values of x, it is necessary to recall the earlier principle: the calculated x value must allow subtraction from the RHS of the equation. When the radius on the RHS is negative, the calculation must move towards zero from the negative direction. Subtracting a positive number from a negative RHS would expand the value away from zero, artificially increasing the size of the radius—which is geometrically impossible. Therefore, the calculated value of x must operate in a direction that reduces the magnitude of the RHS towards zero. The final component is the value of y. This value achieves two objectives: it identifies a specific coordinate on the circumference of the circle, and it determines the geometric length of the vertical side of the triangle by calculating the absolute value (the change in y). The final coordinate for y is not inherently a negative value itself; rather, its position on the grid is determined by its relationship to the centre axis. The radius establishes the negative operational direction, and the y value naturally follows that path. This mirrors the exact logic used when calculating positive values for the upper hemisphere of the circle.

Note that as established in earlier chapters, a multiplier does not alter the inherent sign of the multiplicand. When using the multiplier for graphing the circle, placing it outside of the parentheses does not change the internal value; it is merely describes the direction that the multiplicand is moving. When wanting to find values of the lower part of the circle, apply $^{-}1$ outside of the parentheses as the multiplier. The $^{+}1$ shown outside of the parentheses is still a multiplier and is there simply for clarity and comparison of the negative multiplier. Also keep in mind that the calculated values within the parentheses are a difference, never a negative value:

$$(x-3)^2+(y-3)^2=4$$

For the top half of the circle let $x=2$

$$\begin{aligned}
({}^{+}2-{}^{+}3)^2+(y-{}^{+}3)^2&={}^{+}4\\
{}^{+}1({}^{+}2-{}^{+}3)^2+(y-{}^{+}3)^2&={}^{+}1(4)\\
{}^{+}1({}^{+}1)^2+(y-{}^{+}3)^2&={}^{+}4\\
{}^{+}1({}^{+}1)+(y-{}^{+}3)^2&={}^{+}4\\
{}^{+}1+(y-{}^{+}3)^2&={}^{+}4\\
{}^{+}1-{}^{+}1+(y-{}^{+}3)^2&={}^{+}4-{}^{+}1\\
(y-{}^{+}3)^2&={}^{+}3
\end{aligned}$$

$$y - {}^{+}3 = \sqrt{{}^{+}3}$$
$$y = {}^{+}3 + \sqrt{{}^{+}3}$$
$$y = {}^{+}3 + {}^{+}1.73205$$
$$y = {}^{+}4.73205$$

For the bottom half of the circle let $x = 2$

$$({}^{+}2 - {}^{+}3)^2 + (y - {}^{+}3)^2 = {}^{+}4$$
$${}^{-}1({}^{+}2 - {}^{+}3)^2 + (y - {}^{+}3)^2 = {}^{-}1({}^{+}4)$$
$${}^{-}1({}^{+}1)^2 + (y - {}^{+}3)^2 = {}^{-}4$$
$${}^{-}1({}^{+}1) + (y - {}^{+}3)^2 = {}^{-}4$$
$${}^{-}1 + (y - {}^{+}3)^2 = {}^{-}4$$
$${}^{-}1 - {}^{-}1 + (y - {}^{+}3)^2 = {}^{-}4 - {}^{-}1$$
$$(y - {}^{+}3)^2 = {}^{-}3$$
$$y - {}^{+}3 = \sqrt{{}^{-}3}$$
$$y = {}^{+}3 + \sqrt{{}^{-}3}$$
$$y = {}^{+}3 + {}^{-}1.73205$$
$$y = {}^{+}1.26795$$

Why Does π Appear to Continue Forever?

Traditional textbooks describe circles as having an infinite number of sides. This is generally stated so that the point of tangency can be calculated through calculus, insisting that if an observer zoomed in far enough, a straight side would appear in the graphing environment. While this model is an amazing way to visualise points on a curve, calculus remains perfect whether that straight line is physically there or not.

In reality, a straight line and a curve are fundamentally different physical realities. No matter the magnification, a straight line can never truly become the curved edge of a circle. This distinction leads into the wonder of pi and why it does not end. Searching for an answer online usually results in the claim that pi is irrational—and that is where the explanation typically ends.

However, the concept of pi emerged when a ratio was designed from the properties of a circle. This ratio does something unique: it keeps searching for an end to its own quest. While the idea of straight-sided circles makes calculus visually appealing, pi reacts differently to actual straight edges. If pi interacts with a polygon, the process simply stops because the ratio has finally found a defined side length. In this way, the infinite decimal is not a property of the shape itself, but a mechanical by-product of the ratio used to measure it. Hence, pi is never-ending; because a circle has no straight sides, it possesses only one side that is continually curving.

14

Negative Volume, Displacement and Shapes in General

To understand the final proofs in this book, a common misconception must be challenged: the idea that negative objects actually exist in the physical world. Looking out across the horizon, there is not a single negative tree or a negative mountain to be seen. Even temperature, which is often described as being below zero, is a human-made scale. In the world of science, there is a starting point called absolute zero. At this point, all motion stops and heat ceases to exist. There is nothing colder than absolute zero; therefore, it is the true, physical zero of the universe. When a negative sign is used in mathematics—whether in calculus to describe a ball falling from a roof or in an exponential formula to show a liquid cooling—it is not describing a negative substance. Instead, the sign describes direction and displacement. This distinction is vital. It allows for the utilisation of concepts like negative volume. A negative volume is not a magical box that holds less than nothing; it is a mathematical instruction describing space that has been displaced or removed from a whole. By viewing negative as a description of movement or state-change rather than a physical object, the clutter is removed and the true, logical behaviour of the numbers is revealed. To see the true power of the principles established in earlier chapters, they can now be applied to the physical reality of a 2 metre deep hole. In the standard world of mathematics, a rigid rule persists: a negative multiplied by a negative must result in a positive. However, as already proven, this sign-flipping creates a logical contradiction when measuring the physical world.

The Final Proof: The Negative Volume

If the width of a hole is defined as $^{-}3$ metres and the length as $^{-}5$ metres—representing a removal of space from the surface—the previous proof that a negative multiplied by a negative remains negative in a state of displacement can be utilised. Therefore, the area of the bottom of the hole is correctly identified as $^{-}15$ square metres (m^2). This is a critical moment of clarity. By maintaining the negative identity, there is an acknowledgment that the floor of the hole is still part of the nothingness that has been created. It is an area of displacement, not a positive surface that has somehow reappeared. Following this logic to its conclusion, the area of $^{-}15m^2$ is multiplied by the depth of $^{-}2$ metres. Because the hiccups of traditional rules have been bypassed, the calculation remains consistent, resulting in a total Negative Volume Displacement of $^{-}30$ cubic metres (m^3)

By trusting the foundation built in the first half of this book, the math finally matches the shovel: the negative sign stays attached to the excavation from the first measurement to the last.

15

More New Possibilities: The Real-World Impact

The corrections established in this book are not merely academic. When mathematics aligns with physical reality, the clutter of complex workarounds disappears, revealing a streamlined path forward for science and technology. The following areas represent the potential transformation when the consistent identity of negative values is embraced.

1. A Unified Physics

The subatomic world currently relies on the square root of negative one to function. Equations such as the Schrödinger Equation use these imaginary values to describe the behaviour of matter.

- The Shift: Providing a real solution to negative multiplication removes the need for hidden dimensions or complex phases.
- The Result: This simplifies String Theory and leads to a Theory of Everything calculated on a single, continuous number line. The manipulation of matter and energy becomes far more intuitive.

2. The Next Leap in Computing

Modern wireless technology, including 5G, Wi-Fi, and Satellite communications, treats electricity as a rotation to account for phase shifts. This requires intense processing power to manage complex numbers.

- The Shift: If these shifts are calculated on a one-dimensional real number line, the mathematical load on processors drops significantly.
- The Result: Signal processing becomes exponentially more efficient. This leads to a massive leap in computing power, allowing smaller devices to handle tasks that currently require supercomputers.

3. Precision in Robotics and Spaceflight

To rotate an object in 3D space, engineers currently use Quaternions—complex systems using multiple imaginary units.

- The Shift: A real solution allows these 3D rotations to be calculated using the same basic addition and multiplication used in everyday life.
- The Result: Autonomous navigation becomes more precise and less math-heavy. Robotics becomes cheaper and more accessible, as simpler processors are capable of handling highly complex movements.

4. Advanced Structural Engineering

In traditional engineering, stress and strain are calculated using absolute values or squared terms that often hide the true direction of force.

- The Shift: Allowing negative products to remain negative permits the modelling of compression-on-compression forces with absolute accuracy.
- The Result: Bridges, skyscrapers, and aircraft are designed with higher mechanical truth. Engineers identify exactly

where forces compound in a negative direction, leading to safer structures and materials that thrive under extreme negative pressure.

5. Thermodynamics and Cold-Energy Transfer

Thermodynamics usually treats cold simply as a lack of heat. However, in deep-space applications or cryogenics, cold behaves as an active force.

- The Shift: If multiplying two negative thermal values results in a negative product, a dedicated Cold Algebra is created. This describes how sub-zero energy states interact without forcing them into heat-based models.
- The Result: This revolutionises superconductivity and quantum cooling. More efficient ways to reach absolute zero or manage heat-waste in data centres are found by treating negative energy as a consistent identity.

6. Fluid Dynamics and Turbulence Modelling

Weather systems and ocean currents are modelled using vectors. When two counter-rotating currents meet, math becomes cluttered to avoid losing the sign.

- The Shift: A system where the product of two negative rotations remains negative allows for a one-way calculation of turbulence. It prevents simulations from accidentally flipping the direction of a storm or current.
- The Result: This leads to near-perfect weather forecasting and more efficient hull designs for ships and submarines. The behaviour of complex fluid systems is predicted accurately because the math respects the consistent direction of the flow.

7. Biological Rhythms and Neural Coding

The human brain and heart function through constant, rhythmic oscillations.

- The Shift: Currently, medical researchers use complex Fourier transforms to analyse EKG and EEG data. A negative base logarithm naturally describes systems that oscillate between states or directions.
- The Result: More accurate models for cardiac health and neurological patterns are developed. This allows for smarter pacemakers or neural interfaces that speak the same mathematical language as the human body.

8. Correcting Economic Models

Economic models struggle with market reversals when a trend flips from growth to debt. Traditional math treats these as separate events or shocks.

- The Shift: By using a negative base, growth and contraction are modelled on the same continuous line. Debt is a natural continuation of the pattern rather than a break in the system.
- The Result: This leads to far more accurate predictive modelling for global markets. The onset of a recession or a market bubble is identified earlier by observing the mathematical identity shift towards a negative magnitude.

9. Advanced Materials and Metamaterials

Metamaterials are currently developed with a negative refractive index to bend light in unconventional directions.

- The Shift: These materials are currently designed using complex lateral math to force light to behave in specific ways.
- The Result: If negative multiplication and square roots are real operations, these materials are designed using straightforward, linear calculations. This speeds up the development of perfect lenses, stealth technology, and ultra-efficient solar panels.

10. Artificial Intelligence and Neural Networks

AI learns by weighting connections between nodes. When these weights flip or inhibit a signal, math becomes cluttered with sign-checks.

- The Shift: A number system that handles negative identities naturally allows an AI to process inhibitory signals with the same speed as excitatory signals.
- The Result: This leads to True Symmetrical AI. Neural networks require less code to manage negative inputs, making them faster, more energy-efficient, and closer to the way biological neurons fire.

11. A New Dimension of Data Encoding

Standard logarithms are linear and predictable. A negative base logarithm possesses a natural relationship with direction and identity.

- The Shift: Information is encoded in the rotations or phases of a number rather than just its magnitude.
- The Result: This leads to massive leaps in cryptography and data compression, where a single value holds far more information than previously possible.

Glossary

Natural Order: The mathematical framework that aligns numerical operations with physical reality and the laws of the universe. It removes magic sign-flipping in favour of consistent identity.

Multiplicand (The Identity): In the Natural Order, the multiplicand is the Noun or the primary substance of an equation. Its identity (positive or negative) dictates the identity of the final product.

Area of Displacement: A measurement of a surface that represents the removal of space (such as the floor of a hole). Unlike traditional math, which views all areas as positive, the Natural Order recognises this as a negative surface.

Successor Function (Physical): The logical progression from one unit to the next. In the Natural Order, this function is bound by physical space, meaning the successor must exist as a tangible addition or subtraction in reality.

Absolute Zero: The state of total non-existence or empty space before a unit is introduced. In this book, zero is treated as a physical boundary, not just a placeholder on a line.

Commutative Correction: The rule in the Natural Order that while the *result* of an equation may be the same (2×3 vs. 3×2), the *physical story* is different depending on which number is the Multiplicand (the Boss).

References

- **Australian Government Department of Education (2022).** *Belonging, Being and Becoming: The Early Years Learning Framework for Australia (V2.0).* *

- **Edwards, C., Gandini, L., & Forman, G. (2011).** *The Hundred Languages of Children: The Reggio Emilia Experience in Transformation.* Praeger.

- **Descartes, R. (1637).** *La Géométrie. * (This shows you are engaging with the foundation of the X/Y grid system.)*

- **Euler, L. (1770).** *Vollständige Anleitung zur Algebra (Elements of Algebra).*

- **Gauss, C. F. (1801).** *Disquisitiones Arithmeticae.*

- **Peano, G. (1889).** *Arithmetices principia, nova methodo exposita (The Principles of Arithmetic).*

About the Author

Chris is an Australian early childhood educator with a deep-seated passion for rethinking how we teach the foundations of logic and mathematics. Driven by an innate curiosity from an early age, Chris found himself constantly challenged by the conventional system—not because he couldn't understand mathematics, but because he consistently spotted underlying logical glitches that standard rules failed to explain. This frustration resurfaced years later when trying to guide his own children through the traditional school curriculum. Realising that the system prioritised procedural compliance over physical reality, Chris set out to uncover the truth behind these anomalies, leading to the ground-breaking framework presented in this book.

This book is the result of years of independent research into number theory and physical geometry. By challenging the long-held magic tricks of multiplication and restoring what he calls the Natural Order, Chris has developed a system where maths finally matches the physical world. He openly acknowledges that his greatest fortune in life was meeting his wife, Tracy, and together raising their beautiful family. When he isn't decluttering the number system or researching the Collatz Conjecture, he can be found following the Richmond Tigers or sharing the joy of discovery with Tracy and their children Skye, Willow, and Declan.

To connect with me directly or to share your thoughts on the framework, you can reach out at: chris@thenaturalorder.net.au